M. S Lovell

The Edible Mollusks of Great Britain and Ireland

With Recipes for Cooking Them

M. S Lovell

The Edible Mollusks of Great Britain and Ireland
With Recipes for Cooking Them

ISBN/EAN: 9783744787208

Printed in Europe, USA, Canada, Australia, Japan

Cover: Foto ©Andreas Hilbeck / pixelio.de

More available books at **www.hansebooks.com**

THE

EDIBLE MOLLUSKS

OF

GREAT BRITAIN AND IRELAND

WITH

RECIPES FOR COOKING THEM.

BY

M. S. LOVELL.

"And the recipes and different modes of dressing
I am prepared to teach the world for nothing,
If men are only wise enough to learn."
Athenæus, Deipnos. Bk. iii. c. 69.

LONDON:
REEVE & CO., 5, HENRIETTA STREET, COVENT GARDEN.

1867.

PREFACE.

In these days, when attention has been so much directed towards the cultivation of the common kinds of eatable shell-fish, it is surprising that the importance of certain others for food has been hitherto almost entirely overlooked. We understand the good qualities of oysters, cockles, and a few other kinds; but some equally nutritious (which are universally eaten on the Continent) are seldom, if ever, seen in our markets, or are only used locally as food, and the proper modes of cooking them are scarcely known. I have therefore endeavoured to call attention to all the eatable species common on our coasts, and also to those which, though not found here in abundance, might be cultivated as easily as oysters, and form valuable articles of food.

M. S. Lovell.

PRINTED BY J. E. TAYLOR AND CO.,
LITTLE QUEEN STREET, LINCOLN'S INN FIELDS.

CONTENTS.

	PAGE
HELICIDÆ	1
CARDIADÆ	27
SOLENIDÆ	39
CYPRINIDÆ	43
MYTILIDÆ	45
OSTREADÆ	68
PECTINIDÆ	98
HALIOTIDÆ	114
PATELLIDÆ	119
MURICIDÆ	124
LITTORINIDÆ	136
AVICULIDÆ	139
VENERIDÆ	143
TELLINIDÆ	150
MACTRIDÆ	152
MYADÆ	155
PHOLADIDÆ	157
SEPIADÆ	163
CIDARIDÆ	176
LIST OF WORKS REFERRED TO OR CONSULTED	181
INDEX	189

ILLUSTRATIONS.

PLATE I. (*Frontispiece.*)
    1 & 2. Helix pomatia.—Apple or Vine Snail.
    3 & 4. Helix nemoralis.—Wood Snail.
    5. Helix aspersa.—Common Garden Snail.
    6. Helix Pisana.—Banded Snail.

PLATE II. 1. Cardium edule.—Common Cockle.
    2. Cardium rusticum.—Red-nose Cockle.

PLATE III. 1. Mya truncata.—Gaper.
    2. Solen siliqua.—Razor-shell.

PLATE IV. Isocardia Cor.—Heart-shell, or Oxhorn Cockle.

PLATE V. 1. Mytilus edulis.—Common Mussel.
    2. Ostrea edulis.—Oyster.

PLATE VI. Pinna pectinata.—Sea-wing.

PLATE VII. 1. Tapes pullastra.—Pullet.
    2. Venus verrucosa.—Warty Venus.

PLATE VIII. 1. Psammobia vespertina.—The Setting Sun.
    2. Mactra solida.—Trough-shell.

PLATE IX. 1. Pecten opercularis.—Painted Scallop.
    2. Pecten maximus.—Scallop.

PLATE X. 1. Haliotis tuberculata.—Ormer, or Ear-shell.
    2. Patella vulgata.—Limpet.

PLATE XI. 1. Buccinum undatum.—Whelk.
    2 & 3. Littorina littorea.—Periwinkle.

PLATE XII. Pholas dactylus.—Piddock or Clam.

EDIBLE BRITISH MOLLUSKS.

SNAILS, COCKLES, RAZOR-SHELLS, ETC.

Fam. HELICIDÆ.

HELIX.—SNAIL.

Helix pomatia, Linnæus. *Apple Snail.*—Shell globular, strong, large, covered with coarse longitudinal striæ, 5 volutions, couvex; spire short, and the apex blunt; pale cream-colour, with rufous bands; the columella arched; and of a pale purplish-pink; the outer lip dark reddish-brown; mouth almost round.

Helix aspersa, Linnæus. *Garden Snail.*—Shell opaque, nearly globular, 4 to 4½ volutions, the last much larger, occupying nearly two-thirds of the shell; mouth nearly ovate; spire short, with a blunt point; the outer lid white, with dark-brown bands or mottlings, subject to great variety of markings; epidermis yellowish-green, and thick.

Helix nemoralis, Linnæus. *Wood Snail.*—Shell imperforate, globular, whorls 5, more or less covered with minute spiral striæ; mouth pyriform; inner margin of lip of a rich, dark chocolate-brown; in variety *hortensis* mouth has a white lip. Colours various; yellow, yellow with brown bands, pink, pink and brown, dark chocolate, with darker bands of the same colour, and white.

HELIX PISANA, Linnæus. *The Banded Snail.*—Shell rather depressed and nearly globular, of a pale yellowish-white, with spiral bands of a dark chocolate-brown, which are not always joined together, giving the shell a speckled or streaky appearance; whorls 5 or $5\frac{1}{2}$; mouth pink, and rather large. Varieties nearly white, and also others with the bands of a chestnut-colour, and scarcely to be distinguished.

Helix pomatia is the largest of our land snails, being about $1\frac{3}{4}$ inches in breadth and length, and is found in Kent, Surrey, Gloucestershire, and other southern counties; and a specimen was met with some time since in a lane near Exmouth, which I believe to be a new locality for it. Some curious reversed specimens are occasionally found in France, and one variety particularly struck me, which was exhibited in the Museum at the Jardin des Plantes, in Paris. It was something the shape of a *Buccinum*, the whorls rounded and swollen, and six in number. A beautiful white variety is also found, but rarely, in the environs of Clermont. It is supposed by some to have been originally introduced into England by Sir Kenelm Digby, as food or medicine for his wife, who was suffering from consumption; others say that the Romans introduced it; but Mr. Jeffreys believes it to be indigenous, and observes (in his 'British Conchology') "that it is not found in many parts of England and Wales where the Romans built cities or had important military stations."

Archæologists often find snail-shells in great abundance, however, in excavating on the sites of Roman stations, and at Lymne, in Kent (Portus Lemanis), Mr. Wright has seen them dug up in masses almost as large as ordinary buckets, and completely embedded together.* In France also empty shells of the apple snail have been met with amongst the ruins of Roman villas, in the neighbourhood of Auch, Agen, and in Provence; and in the Danish "kjökkenmöddings," *Helix nemoralis* has been found in small quantities.

As a medicine, snails were recommended for other diseases besides consumption, and *Helix aspersa*, the common garden snail, was generally used.

In a quaint old book, entitled 'A Rich Storehouse or Treasurie for the Diseased, wherein are many approved medecines for divers and sundrie diseases which have been longe hidden, and not come to light before this time; first set forth for the benefit of the poorer sorte of people that are not of abilitie to goe to the Physicians,' by Master Ralph Bower, we find:—" Snales which bee in shells, beat together with bay salt and mallowes, and laid to the bottomes of your feet, and to the wristes of your hands, before the fit commeth, appeaseth the ague." Again :—"Take 20 garden snailes, and beate them (shelles and all) in a morter, untill you perceiue them to be come to a salue; then spread a little thereof upon a linnen cloath, and lay it to the place grieued, and when one plaister is dry, then take that of, and put on another, and it will both heale the sore place and draw it." For corns, he recommends " blacke sope and snailes, of each a like quantitie, stampe them togither, and make a plaister thereof, and spread it upon a piece of fine linnen cloth, or else upon a piece of white

* 'The Celt, the Roman, and the Saxon.'

Leather, and lay it upon the corne, and it will take it cleane away within 7 dayes space.

"*An other soueraigne Medecine for a Web in the eye.*—Take a good quantitie of snailes with their shells upon them, and wash them very well, and then distill them in a common Stillatorie; then take of the galles of Hares, Red Currall, and Sugcr-candie, and mingle them together with the said water, and then distill them againe; then take the same water, and put it into a glasse or viall, and when you will use it, take a drop thereof, and put it into your eyes both morning and evening, and it will help you."

Dr. Fuller, in his 'Pharmacopœia,' recommends snails in scorbutic affections, and gives the following recipe for a consumption:—

"*Snail-water Pectoral.*—Take snails beaten to a mash with their shells 3 pound; crum of white bread, new-baked, 12 ounces; nutmeg, 6 drams; ground-ivy, 6 handfulls; whey, 3 quarts; distill it in a cold still, without burning. If I would have this water not so absolutely cold, I add brandy half a pint or a pint. This water humects, dilutes, supplies, tempers, nourishes, comforts, and therefore is highly conducive in hectic consumptive emaciations."

In Dr. John Quincy's 'Pharmacopœia Officinalis, or a complete English Dispensatory,' are the following:—

"*Decoctum Limacum, or decoction of snails.*—Take garden snails, cleansed from their shells, no. 12; red cows' milk, new, two pounds; boil to a pound; and add rose-water, an ounce; sugar-candy, half-an-ounce.

"It will be very difficult to boil this so long as to waste one-half, because it will be apt both to run over and burn to the bottom, and therefore must be stirred all the while; this quantity is ordered to be drank every morning, and is a noble restorative in consumptions, especially for young people."

"*Decoctum Antiphthisicum, a Decoction against consumptions.*—Take ox-eye daisy flowers, dried, a handful; snails, washed clean, numb. 3; candied eryngo-root, half an ounce; pearl-barley, 2 ounces: boil in spring-water from a pound to half a pound, and then strain for use.

"This brings a supply of such soft and inoffensive nourishment, as gives no trouble to a weak constitution, and therefore is of service in consumptions, hectic fevers, etc. etc. The patient must drink 4 ounces of this warm, with an equal quantity of milk, twice a day."

In Ireland the snail or "shellimidy" was recommended for many diseases, and "a water distilled from shell-snails in canary wine, in the month of May, is a great restorative in consumptions; also strengthens the liver; outwardly applied it is a cosmetic; it beautifies the face, and the volatile oil and spirit extracted from snails resist poison, open all manner of obstructions, cure the pleurisy, asthma, most disorders of the lu gs, and, after a wonderful manner, the consumption. Dose of the volatile salt, from grains 6 to 12; of the spirit from 30 to 40 drops."*

The following recipes are from an old manuscript book; but though snails might be tolerated, I doubt any person having sufficient courage to try them with the addition of earthworms!

"*For a Consumption.*—Take 12 snails, better house snails, and 12 earthworms, clean washed; boil them in a pint of new milk to half a pint, then pour it on 1 ounce of eryngo-root. Take some every night and morning."

* 'Zoologia Medicinalis Hibernica,' by John Keogh.

"*Mock Asses' Milk.*—One pound of snails layed in salt and water for 2 days, and then cleaned and washed, a quarter of a pound of barley, 3 pennyworth of eryngo-root; boil all the above together till they become a jelly, and let them be strained off. Half a pint night and morning for a grown person, and ¼ of a pint for a child. It must be taken warm, and a little milk and sugar added after it is warmed. It is an excellent remedy for consumption and any weakness."

"*For a Consumption.*—24 garden snails, 2 sheeps' trotters, half an ounce of comfrey-root, one quart of spring-water, a quart of milk; boil all together till reduced to half the quantity; take a cup of this night and morning."

"*For a Swelling on the Joints.*—Take 3 handfulls of shell snails (off a rabbit-warren), pound them very fine, and mix them with some new milk (not too thin); put them between two pieces of fine linen cloth, and apply them on the part. This is to be applied once a day, or as often as it gets dry."

A modern authority, Francatelli, gives the following recipe in his 'Cook's Guide':—

Mucilaginous Broth.—Put a cut-up chicken, a pound of veal cutlet, and a calf's foot into a stewpan, with 3 pints of water, boil and skim; then add a dozen crayfish and a pint of *garden snails*, both bruised raw in a mortar; add also a handful of balm, burrage and chervil, 3 ounces of prepared Iceland moss, and a small quantity of salt. The broth must boil very gently by the side of the fire for about two hours without much reduction, and when done is to be strained into a basin for use.

Note.—This is a powerful demulcent, and is much in

use in France in cases of phthisis, catarrh, bronchitis, etc. etc.

Oil of Black Snails—Spanish Cure for Consumption.—Make a flannel bag of a triangular shape (like a jelly bag), fit the corner into a wide-mouthed bottle, fill it with *black snails* in the hottest time of the year; tie up the mouth, and suspend bottle and bag on a wall, the hottest you can find. The proper place is the sunny angle of a wall where the south and west sun fall longest. The snails will give out a larger quantity of frothy liquid, which will drain into the bottle: cork it close for use, and give a teaspoonful at a time, three or four times a day, in milk or any other liquid.

The common garden snail, *Helix aspersa,* also gives out a frothy liquid which might be collected in the same manner, and used with benefit by consumptive patients. The friend who kindly gave me the above recipe tells me that these black snails resemble *H. aspersa,* but the colour is much darker, and at a distance looks almost black. In an old English medical book, date 1756, syrup of snails is recommended for coughs, weaknesses, etc., and is made by hanging snails up in a bag, with some sugar, by which means the syrup drops into a vessel placed to receive it.

Popular Spanish Cure for the Headache.—Make a poultice of bruised snails. They must be broken up with their shells and put into a piece of linen folded 4 times so as to make it thick, dip it in brandy, and squeeze it tolerably dry; then apply it to the forehead.

M. Figuier remembers when studying botany in the garden of the School of Medicine, as a youth, at Montpellier, seeing the celebrated tenor singer, M. Laborde, every morning partake of live snails, as he was suffering from a weak chest. M. Figuier assisted in finding the snails in the holes in the garden wall, and under leaves, and M. Laborde crushed the mollusks with a stone, picking off the pieces of broken shells, then rolling the fish in powdered sugar swallowed them. The remedy was evidently efficacious, as twenty years later M. Laborde still held his position as tenor, and sang at the theatre at Brussels and also at the opera in Paris.*

In the 'Meddygon Myddvai,' published by the Welsh MSS. Society, the following recipes are found:—

For an Impostume (whitlow).—Take a snail out of its shell, and bruising it small, pound it into a plaster and apply it to the finger; it will ripen and break it, and it should then be dressed like any other wound. For "a patient who is burnt" it recommends a plaster of mallows, snail-shells, pennywort, and linseed pounded, and applied until the part is healed without even uncovering it.

In olden times it was supposed that the small grits of sand found in the horns of snails, introduced into hollow teeth, removed the pain instantaneously; and that the ashes of empty snail-shells mixed with myrrh were good for the gums. (Pliny's Nat. Hist. vol. v. p. 431.)†

Pliny also recommends "snails beaten up raw and taken in 3 cyathi of warm water for a cough," and a snail diet for internal pains, the snails to be cooked as follows:—"They must first be left to simmer in water for some time without touching the contents of the shell; after which, without any other addition, they must be grilled upon hot coals, and eaten with wine and garum" (chap. 15, book xxx.), (a kind of fish-sauce).

* 'La Vie et les Mœurs des Animaux,' p. 386.
† Throughout this volume I have used the translations of Pliny and Athenæus, in Bohn's series of Classical Authors.

Again, that a kind of small, white, elongated snail, dried upon tiles in the sun and reduced to powder, then mixed with bean-meal in equal proportions, forms a cosmetic for whitening and softening the skin.

I have been told that a large trade in snails is carried on for Covent Garden market in the Lincolnshire Fens, and that they are sold at 6*d*. per quart, and upon further inquiry I find that snails are still much used for consumptive patients and weakly children; also as salves for corns put between ivy leaves; and as food for birds. In the manufacture of "*cream*" they are also much employed, bruised in milk and boiled, and a "*retired*" milkman pronounced it the most successful imitation known.

It appears that not only are the *Helicidæ* nourishing to the human species, but that they have a beneficial effect upon sheep, giving a richness to the flavour of the mutton. Mr. Jeffreys, in his 'British Conchology,' quotes the following passage from Borlase's 'Natural History of Cornwall:'—"The sweetest mutton is reckoned to be that of the smallest sheep, which usually feed on the commons where the sands are scarcely covered with the green-sod, and the grass exceedingly short; such are the towens or sand-hillocks in Piran-sand, Gwythian, Philne, and Senan Green, near the Land's End, and elsewhere in like situations. From these sands come forth snails of the turbinated kind, but of different species, and all sizes, from the adult to the smallest just from the egg; these spread themselves over the plains early in the morning, and whilst they are in quest of their own food among the dews, yield a most fattening nourishment to the sheep."

Birds also are great eaters of snails. Lister mentions

the partiality of thrushes for *Helix nemoralis*; and owing to the scarcity of this species in South Derbyshire, I have twice brought a large basketful of live specimens from Staffordshire, and turned them out, hoping they would thrive and increase; but I have not only found the dead and broken shells, but constantly disturbed the feathered depredators themselves at their repast. *Helix arbustorum* I have also tried, but with the same success; they fared no better than the other kind.

There is a true saying "that there is nothing on earth so small that it may not produce great things."* Thus, the sacred geese at Rome by their cackling awoke Marcus Manlius, and thereby saved the Capitol from the Gauls, who were attempting by night to surprise the garrison; and even such insignificant creatures as snails were the cause of the following disaster to a Numidian king:—A castle on a lofty and steep rock, into which Jugurtha had carried all his treasures, had long been besieged in vain by Marius, when a Ligurian in the Roman army, climbing up the rocks in quest of snails, was led to continue his search for them, till he had nearly reached the summit, and thus found that the ascent was practicable; and on reporting this fact to Marius, having been ordered to lead a chosen band up the same part of the rocks, he and his comrades so alarmed the garrison by their unexpected appearance that they gave up the castle to the besiegers.

The Romans were very partial to snails as an article of food, and fed them till they grew to a large size. Several sorts are mentioned by Pliny, and they were all kept separate; amongst others, white ones that were found in the neighbourhood of Rieti. He describes the Illyrian

* Proverbial Philosophy.

snails as the largest, the African as the most prolific; others from Soletum, in the Neapolitan territory, as the noblest and best. He also speaks of some as attaining to so enormous a size that their shells would contain 80 pieces of money of the common currency,* that is to say, 80 quadrantes, the quadrans being a small copper coin $\frac{3}{4}$ of an inch in diameter, about the size of a new sixpence, and $\frac{1}{16}$ of an inch thick. This statement of Pliny's is really not so improbable as may appear at first sight, for on trying how many sixpences a usual-sized specimen of our largest snail, *Helix pomatia*, would hold, I find that about 40 could easily be put into it; and in the museum of the Jardin des Plantes, in Paris, there are two specimens of this *Helix* from Moldavia, nearly twice the size of the usual ones, measuring about $2\frac{1}{4}$ inches in breadth, and which would easily hold 80 sixpences.

Fulvius Hirpinus studied the art of fattening them with so much success, that some of his snails would contain about 10 quarts. Pliny, in his letter to Sextus Erucius Clarus, says (complaining of his not fulfilling his engagement to sup with him) :—" I had prepared, you must know, a lettuce apiece, three snails, two eggs and a barley cake, with some sweet wine and snow."†

In Sir Gardner Wilkinson's 'Dalmatia and Montenegro,' he tells us that the Illyrian snails mentioned by Pliny‡ are very numerous in Veglia or Veggia, the Cyractica of Strabo.

Both *Helix pomatia* and *Helix aspersa* are eaten abroad to this day, and in England Dr. Gray says§ that the glass-

---

* Kirby's Hist. of Animals, etc., 'Bridgewater Treatise,' vol. i. p. 284.
† Pliny's Letters, p. 30, vol. i.   ‡ Pliny, ix. 56.
§ Blackwood's Edin. Mag., no. 561, July, 1862.

men at Newcastle indulge themselves in a snail-feast once a year, and collect them from the fields and hedgerows on the Sunday previous.* Addison, in his Travels, mentions having seen a snail garden, or " escargotière," at the Capucins, in Friburg. It was a square place boarded in, filled with a vast quantity of large snails. The floor was strewn about half a foot deep with several kinds of plants, for the snails to nestle amongst during the winter. When Lent arrived, their magazines were opened, and a ragoût made of snails.

"Escargotières," or snail gardens, have been in use for a length of time in various parts of Europe. Dr. Ebrard, in his pamphlet, 'Des Escargots,' states that M. Fischer, of Bordeaux, mentions those of Brunswick, Silesia, and Copenhagen, which latter furnished snails for the tables of the noble Danes in the eighteenth century.

In Barrois, an "escargotière" consists of a cask with the head staved in, covered with a net; or a square hole with the sides lined with wood, and fastened over at the top with an iron trellis, or with a simple hurdle made of light osier-sticks. The snails are placed in as they find them, until there are sufficient for a repast, or for sale. They are also kept in these places till they are fattened, or till they close their shells with their epiphragm, which enables them to be more easily transported. In Lorraine, a corner of the garden is often given up to the snails, surrounded with a fine trellis-work to prevent their escaping, and all kinds of vegetables are placed inside which are most appreciated by them.

During the winter, the "escargots" (their shells

* Since the above was written, a correspondent of 'Land and Water' says the glassmen's snail feast, at Newcastle, no longer exists.

being closed with their epiphragm) are kept in pots, jars, or baskets, in a dry cold place. The vine-growers in the neighbourhood of Dijon keep them in a dry cellar, or else dig a trench in the vine-slopes, and place at the bottom some leaves, and then their snails, covering them with more leaves and a few spadefuls of earth.

In Silesia, the snails are fed with marjoram, wild thyme, and aromatic plants, to give them a flavour.

Ulm, in Würtemberg, is celebrated for its "escargotières," and, according to Marteni, "more than 10 millions of *Helix pomatia* are sent away to different gardens and "escargotières" to fatten, and when ready for table are sent to various convents in Austria for consumption during Lent.

Helicidæ are considered rather poor food, and therefore suitable as Lenten fare; and this peculiarity has given rise to a singular custom near Bordeaux, mentioned by M. Fischer, who tells us that every year crowds of people direct their steps towards the township of Canderan, to end the Carnival with gaiety, and to have a foretaste of Lent by feasting on snails. The consumption is considerable, and a dish of 25 snails costs 1 franc 50 centimes.

A friend told me he had often seen the large apple-snail on the dinner table at Vienna; they were served up plain, boiled in their shells, or stuffed with forcemeat. At Naples, snails are generally kept in bran for a week or two, or for two or three days, before they are considered good for the table. They live on the bran, which is said to fatten them.

When first the snails are gathered from the hedges, etc., it is a necessary precaution to starve them for a few days, and not to eat them at once, as they feed on poisonous plants, such as the deadly nightshade, poppy, datura, black nightshade (*Solanum nigrum*), etc., cases of poisoning by snails having occurred where they had been gathered near, or had fed upon these noxious plants.

It is a mistake to suppose that the only snails used as food are the *Helix pomatia* and *Helix aspersa*.* These are naturally preferred on account of their larger size, which makes them less troublesome to eat; but a variety of small kinds of snails, fifteen species in all, including those above mentioned, are also employed in cookery on the Continent, and there is no reason why they should not be as good as the others, nor is there any reason why we should not use snails, and many other molluscous animals, which we now throw aside, but which are doubtless quite as palatable and as wholesome as other kinds which our prejudices permit us to indulge in.

M. le Docteur Ébrard, in his 'Des Escargots, au point de vue de l'Alimentation, de la Viticulture et de l'Horticulture,' gives an interesting account of the use of snails both for food and medicine, and he tells us that during a sojourn of some weeks at Hyères, in the month of April, he was struck by seeing suspended at the side of the door of each cottage, a rush basket of a peculiar form. He was curious to find out the contents, and on looking into one he found it full of snails.

At the sight of these creatures he made a slight movement of disdain, which was perceived by the master of the house, who said, "These snails disgust you, but we

---

* *Helix aspersa* has a variety of names in France, and in the north it is called *colimaçon, jardinière*, and *aspergille;* at Montpellier, *caraguolo*; in Bordelais, *cagouille, limaou*, and *limal;* in Provence, *escargot* and *escourgol;* at Avignon, *caragoou* and *contar;* *banarut* at Arles; and *bajaina* at Grasse.—*Dr. Ébrard.*

poor country people eat no other meat all the year, except at Easter."

Dr. Ébrard adds that, during the famine of 1816 and 1817, snails were most valuable articles of food to the inhabitants of Central France; again, that from the coasts of Saintonge and Aunis, snails have been for a long time exported in casks to Senegal and the Antilles; and that M. Valmont-Bomard saw the peasants, in the neighbourhood of La Rochelle, gathering an immense quantity of *small* snails, to send to America, in casks filled with branches of trees, crossed again and again, so that the snails might be able to attach themselves firmly, and not be much shaken during the transport.

Helix aperta, which is not known in England, but is figured in Messrs. Forbes and Hanley's 'British Conchology,' from a dead specimen having been found in Guernsey, in 1839, is highly esteemed amongst real connoisseurs of snails, and is found in Provence (where it is called by the Provençaux *tapada*, *tapa*, or *tapet*), in some parts of Italy, and in the islands of the Mediterranean.

M. Moquin-Tandon tells us that vessels regularly visited the coasts of Liguria, in search of considerable quantities of *Helix aperta*, for food for the higher classes at Rome.* The shell is of a yellowish-olive colour and nearly transluecnt, thin, and of an ovate-globular form. It has a large mouth, with the peristome white, and the whorls four in number. In the heat of summer and during the winter this *Helix*, like *Helix pomatia*, buries itself in holes in the ground, shutting up the aperture of its shell with a white calcareous epiphragm. Two of the specimens we have in our collection, which were sent from Italy, still have this epiphragm very perfectly preserved, and it

* At Rome, *Helix aperta* is called *Monacello*.

is very glossy and slightly convex. Theophrastus, in his treatise upon animals which live in holes, states that snails have the habit of burying themselves. He says: —"Snails live in holes during the winter, and still more in the summer, on which account they are seen in the greatest numbers during the autumn rains. But their holes in the summer are made in the ground, and in the trees."*

Helix nemoralis is also eaten, and at Toulouse sells for 5 or 10 centimes a dish; but by some, snails with striped shells are not considered good, as they have a bad taste and smell. M. Moquin-Tandon purchased, in 1847, in the market at Toulouse, a basket containing 400 specimens of *Helix aspersa* for 60 centimes; and another, with 1503 specimens of *Helix nemoralis*, for 75 centimes,—making 15 centimes the 100 for the former, and a little less than 5 centimes for the latter. *Helix nemoralis* and *Helix hortensis* are known by various names in France; "for instance, at Bordeaux they are called *demoiselles*, *mogne* at Libournes, *molimorno* at Limoges, *limaio* at Agen, *limaia* at Montpellier, *livrée* in the north of France, and *caracolo* in the Pyrenees."†

Helix Pisana, which is a very local species with us, and only found at Tenby (where I have seen it in profusion), at Manorbeer, in Cornwall, Jersey, and Ireland, is greatly prized as an article of food abroad, and is larger than it is with us,—indeed, almost as large as *Helix nemoralis*.

"At Marseilles, the average sale of *Helix Pisana* and *Helix rhodostoma* is about 20,000 kilogrammes, at 3 francs the 50 kilogrammes, which makes the sum of 1200 francs. By the sale of our common garden snail

---

* Athenæus, Deipn. vol. i. p. 104.   † Dr. Ébrard, 'Des Escargots.'

(*Helix aspersa*) the same price is realized, and that of *Helix vermiculata** amounts to 4800 francs. It is also stated, that in the market at Dijon is sold, annually, about 6000 francs' worth of the vine snail, *Helix pomatia* (the *escargot par excellence*, and called also *luma, gros luma*, and *le moucle dc vigne*), at 1 franc 50 centimes per hundred."†

In Corsica the same species are eaten; and it is said that, in the island of Ré the sale of these *Helicidæ* amounts annually to 25,000 francs, but probably this sum is exaggerated. In Burgundy, Champagne, and Franche-Comté, a great quantity of snails of all kinds are consumed, and also sent to Paris; and Professor Simmonds mentions that there are now 50 restaurants, and more than 1200 private tables, in that city, where snails are considered a delicacy by from 8000 to 10,000 consumers; that the monthly consumption of this mollusk is estimated at half a million; again, that the market price of the vineyard snail (apple or vine snail, *Helix pomatia*) is from 2*s*. to 3*s*. per hundred, while those of the hedges, woods, and forests bring only 1*s*. 6*d*. to 2*s*. He further adds, that in the vicinity of Dijon the proprietor of one snailery is said to clear nearly £300 a year by his snails; and also that there are exported from Crete annually about 20,000 okes (each nearly 3 lb.) of snails, valued at 15,000 Turkish piastres.

M. Renou, (as quoted by M. Cailliaud, of Nantes,) in a curious account, read in 1864 before the Academical Society at Nantes, on the importance that the ancients attached to snails, observed, that during 1862 and 1863

---

* *Helix vermiculata* is sold at Leghorn under the name of *chiocciola*, and at Naples shares in common with other snails that of *maruzze*.

† Dr. Ébrard.

the *escargots* brought to the Marché de la Bourse, at Nantes, on Sundays and *fête* days, amounted in number to 996,000, producing the sum of 2490 francs.*

We read that formerly, in Paris, snails were only to be found in the herbalists' shops and at the chemists'; but now there is a special place for them in the fish market, by the side of the crayfish and other freshwater fishes; and in nearly all the restaurants you may see dishes of *Helix pomatia* displayed in the windows. They are ready cooked, and only require warming for a few minutes on the gridiron. It is from Troyes, at the price of five francs the hundred, that the apple or vine snail is sent to Paris, boiled in their shells, and seasoned with fresh butter mixed with parsley. When you wish to partake of them, you place them before the fire till the butter melts, and then they are fit to eat. I purchased some, and succeeded in eating two, but with difficulty, as the way they were dressed did not disguise the slimy, soapy taste, and the want of salt, pepper, etc., made them most unpalatable. I felt that I could sympathize with Dr. Black and Dr. Hutton, who also endeavoured to eat a dish of stewed snails; but, after vainly attempting to swallow in very small quantities the mess which each internally loathed, "Dr. Black *at length* 'showed the white feather,' but in a very delicate manner, as if to sound the opinion of his messmate. 'Doctor,' he said, in his precise and quiet manner, ' Doctor, do you not think that they taste a little—a very little—green?' 'Green! green, indeed! Take them awa'! take them awa'!" vociferated Dr. Hutton,

---

* 'Catalogue des Radiaires, des Annélides, des Cirrhipèdes, et des Mollusques marins, terrestres et fluviatiles, recueillis dans le département de la Loire-Inférieure,' par Frédéric Cailliaud, de Nantes, p. 222.

starting from the table, and giving full vent to his feelings of abhorrence."*

In Paris, snails are not considered in season till the first frost, about the end of October or beginning of November, when they are closed with their white epiphragm. The Parisians eat about fifteen or twenty for breakfast, and they are also said to give a better flavour to wine.

In Spain, also, all snails are eaten, unless they are too small to cook; and they are called *caracola*, and the men who gather and sell them are called *caracoleros*. However, they apply the term *caracola*, to all snail-like shells, only distingishing them thus, *caracola del mar*, *caracola del rio, caracola del huerta, i. e.* salt, freshwater, or garden caracoles.

Rossmässler mentions having seen fourteen different species of *Helicidæ* brought to the markets in Murcia and Valencia, and sold to be eaten. He adds that snails are not only food for the poor, for that many kinds are too costly. One species, called *serranos*, is sold for a penny of our English money each; but they are not half that price bought by the dozen. They cook them by stewing them, shells and all, in a richly-spiced sauce, and then put the shell to the mouth, and draw out the animal by sipping or sucking it.

Rossmässler states, for the benefit of those who may travel in Spain for scientific purposes, that to collect *plants* it is useless to visit the north of Spain before the middle of April, and the south before the end of March. For *insects* and *shells*, the end of the summer, and above all the autumn, is the best time of the year.

The snail-hunters, who daily supply the markets with

* Curiosities of Food, p. 348.

large baskets of snails, often have to traverse great tracts of hilly country, and are obliged to go out very early in the morning, before sunrise, in search of these creatures, as they are then to be found in more abundance. Much amusement was afforded to the Spaniards, by Rossmässler throwing away the delicate animal, and only retaining its shell, which to them was worthless, but most valuable to him as a conchologist. Upon one occasion, on arriving at a *posada*, he found the hotel people sitting down to their midday meal, before a great dish full of snails. He says:—"One look satisfied me that they were of a rare kind, for which I had sought in vain; and I immediately seized upon some of the empty shells, which caused a universal laugh. I did not care at all for this, but I had actually to pay a real (about 2*s*. 4*d*.) for the empty shells, which, when living, I could have got for nothing." This was thoroughly Spanish.

Dr. W. Gottlob Rosenhauer, in his 'Die Thiere Andalusiens,' says that *Helix lactea*, which is very abundant, and readily found close to stones amongst grass, near Malaga, and San Fernando, is brought in great numbers to the markets in Andalusia, and that the empty shells may be seen there all about the streets. Both *Helix aspersa* and *Helix lactea* are used abundantly for food, but the latter tastes better, and is more delicate. They are generally cooked in rice, with butter or some other greasy substance, and held in a napkin whilst the animal is picked out with a pin; or sometimes the mouth (or head) is first cut off, and the animal is then drawn out by suction,—a proceeding not very elegant, at least according to our English ideas. *Helix lactea* may also be classed among the edible snails of France, and is found in the Pyrenees, and also in Corsica.

Dr. Ébrard was informed by Dr. Roi, the Inspector of Colonization in Africa, that in the market at Algiers large heaps of snails are to be seen of the same species as those found in Central France, and are sold by the bushel, and by the hundred, as an article of food; and a small species, about the size of a pea, is collected in Algeria in great numbers, and given to the ducks.

Sir Gardner Wilkinson has seen baskets full of snails carried about for sale in the streets of Cairo; and in 'Physical Geography of the Holy Land' it is stated that they are occasionally eaten in Syria, though not often. In Scotland, fortunes are predicted by snails. In Hone's 'Every-day Book,' we read that "No one will marry in May, but, on the first morning of that month, the maidens rise early to gather May-dew, which they throw over their shoulder in order to propitiate fate in alotting them a good husband. If they can succeed, by the way, in catching a *snail* by the horns, and throwing it over their shoulder, it is an omen of good luck; and, if it is placed on a slate, then likewise it will describe by its turning the initials of their future husband's name."*

According to the 'Archæologia Cambrensis,' in the parish of St. Clear's, Carmarthenshire, small portions of lands were formerly gambled away by means of snail-

---

* In 'Folklore of the Northern Counties of England,' p. 86, it is said that if, on leaving your house, you see a black snail (slug?), seize it boldly by one of its horns and throw it over your left shoulder; you may then go on your way prosperously; but if you fling it over the right shoulder, you will draw down ill luck. This practice is said to extend as far south as Lancashire.

races. The rival snails were placed at the foot of a post, and the one that first reached the top, won the land for its master. In the Isle of Wight, the fishermen of Atherfield and Brixton consider snails the best bait for prawns, and horseflesh next.

The shells of *Helix pomatia* are used for making small whistles for children. The apex of the shell is cut off, and a piece of tin added; they are then sold for a penny each; and who does not recollect the wonderful cats made of the shells of the common garden snail, *Helix aspersa*, with heads of cement or putty, and how anxious we were to become possessors of these beautiful creatures! They are now seldom seen, except in some small out-of-the-way shop in a country town or village,—such trifles not suiting the tastes of the precocious juveniles of the present day.

The ancients seem to have studied the habits of these mollusks, as besides Theophrastus, whom I have already quoted, Aristotle also mentions them; and Teucer speaks of the snail as "an animal destitute of feet and spine and bone, whose back is clad with horny shell, with long projecting and retreating eyes,"[*] and many others. Hesiod calls the snail the "hero that carries his house on his back," and Anaxilas says—

> "You are e'en more distrustful than a snail,
> Who fears to leave even his house behind him."[†]

Somewhat different is the old English proverbial rhyme,

> "Good wives to snails should be akin,
> Always to keep their homes within;
> Yet unlike snails they should not pack
> All they are worth upon their back."

[*] Athenæus, 'Deipnosophists,' book x. c. 83, p. 720.
[†] Ibid., book ii. c. 63, p. 104.

Gwillim, in his 'Heraldry,' informs us that the snail is called *Tardigrada domiporta*, the "slow-going house-bearer," and adds, "the bearing of the snail doth signify, that much deliberation must be used in matters of great difficulty and importance; for, although the snail goeth most slowly, yet, in time she ascendeth to the top of the highest tower, as Mr. Carew, of Antony, hath wittily moralized in his poem, intituled 'The Herring's Tail.'" He gives snails as the armorial bearings of the Shelleys, but he also mentions whelks, which shells are now borne by this family.

The crest of the Carpenters of Somersetshire is a snail passant proper, shell argent; and that of the Galay family, a snail, horns erect, proper.

To Dress Snails.—Snails that feed on vines are considered the best. Put some water into a saucepan, and when it begins to boil, throw in the snails, and let them boil a quarter of an hour; then take them out of their shells; wash them several times, taking great pains to cleanse them thoroughly, place them in clean water, and boil them again for a quarter of an hour; then take them out, rinse them, dry them, and place them with a little butter in a frying-pan, and fry them gently for a few minutes, sufficient to brown them; then serve with some piquante sauce.*

Snails cooked in the French way.—Crack the shells and throw them into boiling water, with a little salt and herbs, sufficient to make the whole savoury; in a quarter of an hour take them out, pick the snails from the shells, and boil them again; then put them into a saucepan, with butter, parsley, a clove of garlic, pepper, thyme, a bay-leaf, and a little flour; when sufficiently

* An old French recipe.

done, add the yolk of an egg, well beaten, and the juice of a lemon, or some vinegar.

The following are Spanish recipes for cooking them:—

Snails with Parsley—Caracoles con Perejil.—Take a slice of crumb of bread, soak it in vinegar and water, pound it in a mortar with garlic, salt and pepper, parsley and mint; add oil drop by drop, turning the pestle the whole time in the same direction; put the snails which have been already boiled, and taken out of their shells into this, and either serve cold or fry the whole together.

Ragoût of Snails—Guisado de Caracoles.—Soak the snails in salt water, then wash them in two or three waters; take thyme, marjoram, bay-leaves, and salt, and fry them with chopped onions in butter or oil; boil the snails, and take them out of their shells, or, if you prefer it, put them, shells and all, into the butter, and fry them. Let them be served as follows:—Soak a piece of bread in vinegar and water, and pound it in a mortar with a clove of garlic, a little pepper, salt, parsley, and mint, chopped very fine; add oil drop by drop, turning the pestle all the time till it is quite a smooth paste, and place it round the dish, putting the snails in the centre.

Winter Soup.—Place the snails in boiling water for a few minutes, when they will easily come out of the shell. A little bit of hard matter is to be taken from the head, then stew them for a long time in milk.*

Another Recipe from the same source.—Scald the snails to get rid of their shells, and then fry them with a few crumbs of bread, and a little seasoning, viz. pepper, salt, and a finish of fine herbs, or stew them with white or brown sauce.†

* 'Life in Normandy,' vol. ii. p. 24. † Vol. ii. p. 62.

Another French Recipe for dressing Snails.—In spring and autumn, the snails which are found in the vineyards are good to eat, for those who like them; and to clean them and make them easy to get out of the shell they must be dressed as follows:—Take a handful of charcoal ashes, and put it into a saucepan or kettle with some soft water, or water from a river; when it boils, throw in the snails, and leave them for a quarter of an hour. When you find the snails can easily be picked out of the shell, take them and place them in some tepid water to cleanse them; then, again, put them into fresh water, and let them boil for a minute or so, take them out, and let them drain. Put into a saucepan a piece of butter, with a bunch of parsley, chives, a clove of garlic, two cloves, thyme, a bay-leaf, basil, and some mushrooms then add the snails, being careful that they are well drained. Pass the whole over the fire, adding a little flour moistened with broth, a glass of white wine, salt, and pepper, and let it simmer till the snails are quite tender, and till the sauce is nearly dried up in the pan. Serve them up with a sauce made as follows:—Take the yolks of three eggs, beat them up with some cream, warm it, but do not let it boil, add a little white vinegar or verjuice, with a little nutmeg.*

Dijon method of cooking Snails.—Boil them in water with some thyme; take them out of their shells; place in the shells some fresh butter, kneaded with chopped parsley; replace the animal in its shell, and cover it with some more of the butter, etc. When required for eating, place them on an iron dish, or on one of porcelain. They are placed side by side, with the mouth of the shell upwards, in little holes in the iron or porcelain

* 'La Cuisinière Bourgeoise.'

dish, which is made for the purpose, and they must be warmed till the butter melts. Thus prepared, snails sell at Dijon from 5 to 10 centimes apiece.*

Another method of cooking Snails.—In the north and east of France, *Helix pomatia*, or *Hélices vigneronnes*, the apple or vine snails, are boiled in water and taken out of their shells, then stewed in a saucepan with some fresh butter and parsley; or else the snails, after they have been taken out of their shells, and are three parts cooked, are put into a saucepan with a little water and some butter, or with some broth, adding a little salt, pepper, white wine, or vinegar. When they are cooked and tender, pour over them a thickening of yolks of eggs, with chopped parsley; the addition of nutmeg and lemon-juice makes them more savoury.*

The inhabitants of Central France use several sauces for snails, and the four principal are the following, according to Dr. Ébrard, viz.:—

L'ayoli, or *ail-y-oli*, of Languedoc; a paste made with olive oil, and pounded garlic.

L'aillado, of Gascony; a most complicated sauce of garlic, onions, chives, leeks, parsley, etc., with spices, cloves, and nutmeg, the whole thickened with oil.

La limassade, of Provence, called *La vinaigrette* in Paris.

La cacalaousada, of Montpellier, composed of flour, ham, sugar, etc. At Bordeaux the *aillada* is softened with a mixture of bread, flour, and yolk of egg, boiled with milk.

Stuffed snails are also considered very good. A fine stuffing is made with snails previously cooked, fillets of anchovies, nutmeg, spice, fine herbs, and a liaison of

* Dr. Ébrard.

yolk of eggs. The snail-shells are filled with this stuffing, then placed before the fire, and served *very* hot. In some countries Blainville states, that snails are eaten, smoked and dried.

Fam. CARDIADÆ.

CARDIUM.—COCKLE.

CARDIUM EDULE, Linnæus. *Common Cockle.*—Shell equivalve, subcordate, with twenty-four or more ribs radiating from the beaks, which are bent inwards; umbones prominent; the internal margins of the valves fluted or indented. Ligament external, strong, and of a dark horn-colour. Four teeth in each valve; the two primary teeth close together, the lateral teeth remote. Colour yellowish-white.

The common Cockle (the *ruocane* or *bruvane* of the Irish; *la bucarde, sourdon, rigardot,* or *coque* of the French, the *berizon* of the Spaniards) is found all round our coasts, burying itself in sand, or sandy mud, in the neighbourhood of estuaries; and at low tides numbers of people may be seen busily engaged filling their baskets, as it is everywhere much sought after for food; and during times of scarcity in some of the northern islands of Scotland, the inhabitants might have perished with hunger, if it had not been for this useful little shellfish. The quantity of shellfish, particularly of cockles, on the shores of most parts of the Long Island (Western Isles) is almost inconceivable. On the sands of Barra alone, scores of horse-loads may be taken at a single tide. Cockles are considered by the people very nutritious, especially when boiled with milk.* It is astonishing how quickly an expert cockle-gatherer will fill his basket; and sometimes they make use of a piece of bent iron, or half an old hoop, to scrape the shells out of the sands. At Starcross, they have small "cockle-gardens," where the shellfish are kept; and the flavour of these cockles is considered superior to those which are found elsewhere. The costume of the women who gather them is anything but becoming;—large fishermen's boots, their dresses so arranged as to resemble very large knickerbockers, and an old hat or handkerchief on their heads, with their baskets on their backs.

I am told that some of the Gower people, on the *north* side of the seigniory of Gower (a Flemish colony in Glamorganshire), live nine months in the year on cockles. They also carry large quantities to Swansea market, whence they are sent to London.

Mr. Baines, in his 'Explorations in South-West Africa,' tells us that cockle-shells are greatly prized by the Damaras, and, if they are rich enough to afford it, one is worn in the hair over the centre of the forehead; and he adds, that if some friend at home would invest three-halfpence in these favourite mollusks, and send him the shells after his meal, he might make his fortune. In the British Museum a fishing-net is exhibited, from the Friendly Islands, with cockle-shells fastened on it to sink it, instead of leads. Cockle-shells are also used for making garden walks, and good lime is made from them when they are calcined.

In the heraldry of Prussia the cockle-shell is used. "Barry of four, argent and azure, semée of cockle-

* 'Visits to the Seacoasts: Shipwrecked Mariners,' vol. xii. p. 32, 1865.

shells counterchanged, are borne by the Silesian family of Von Strachwitz, which has for crest two wings also charged with cockles."*

We also find this shell figured on coins. A specimen in the British Museum of the *sextans*, the sixth part of the *as*, or piece of two ounces, has on one side a caduceus, a strigil, and two balls, and on the other a cockleshell.

Ossian, in his poem the 'War of Inis-thona,' tells us that the king of that island gave a feast to Oscar, which lasted three days, and that they "rejoiced in the shell,"—meaning that they feasted sumptuously and drank freely. Again, we meet with the "chief of shells," and the "halls of shells." Macpherson calls the cockle the "heroes' cup of festivity," being known by the name of *sliga-crechin,* or the drinking-shell; and it is also stated that this shell is used in the Hebrides for skimming milk.† This seems, however, hardly possible, for the "heroes" would probably not be content with so small a cup as the little common cockle. It must have been some larger shell, and formerly the word "cockle" was applied to *any* shell; besides which, the common cockle could not, from its shape, be used for skimming milk, and from its size, it would be of little use for that purpose. Moreover, we *know* that the so-called cockle used in the Hebrides for that purpose is a *Mya*, there called the cockle.

The Irish, the South Welsh, and probably others, call the whelk (*Buccinum undatum*) the *goggle,* and know it by no other name. It is evidently the same word and is more correctly applied, as we shall presently see.

* Sibmacher's 'Wapenbuch,' Heraldry of Fish, p. 226.
† 'A Book for the Seaside.'

"Cockle" was the common name in olden times for the escallop of pilgrims,—"he wore the cockle in his hat," etc.; and it is still often so used in heraldic language. Lydgate, when he says—

> "And as the *cockille*, with heavenly dewe
> So clene
> Of kynde, engendreth white perlis rounde."

means evidently the *oyster*, alluding to the old fable of pearls being formed by the oyster's rising to the surface at the full moon, and opening its shell to receive the falling dew-drops, which thus hardened into pearls,— an idea which is quaintly detailed by Robinson, in his 'Essay towards a Natural History of Westmoreland and Cumberland' (1709), who, in speaking of the pearls procured from the rivers Irt and End, says, "Those large shellfish which we call horse-mussels, which, gaping eagerly and sucking in their dewy streams, conceive and bring forth great plenty of them" (the pearls), "which the neighbourhood gather up at low-water, and sell at all prices." The natives of India have a similar belief with regard to the origin of pearls, viz. that they are congealed dewdrops, which Buddha in certain months showers upon the earth, when they are caught by the oysters whilst floating on the waters to breathe.*

The natives of Java have a still stranger belief that the *pearls themselves* breed and increase if placed in cotton, and they actually sell what they term "breeding pearls" for this purpose, affecting to distinguish the male from the female. Those pearls which are clustered together, in the form of a blackberry, are said by them to be thus produced. Nor is this belief pecu-

---

* 'Household Words,' vol. iii. p. 80, "My Pearl-fishing Expedition."

liar to Java, as a Spanish lady informed a friend of mine that, if seed-pearls were shut up in cotton-wool, they would increase *either in size or in number!* The experience of our jewellers is that the effect of cotton-wool on pearls is to injure their colour, and make them yellow.

Shakespeare says :—

"Love's feeling is more soft and sensitive
Than are the horns of *cockled* snails."

Here *cockled* means either *shelled* or *whorled*.

The Greek κοχλίας, κόχλος, means a snail, or a shell with a spiral whorl (hence the name of "goggle" for the *Buccinum*); but it is also used sometimes for a bivalve shell or "cockle." Κοχλιάριον is a spoon.

Camden, in his 'Britannia' (p. 962), in speaking of Ireland, and of the commodities of the British Ocean, says :—" There are *cockles*, also, in great numbers, with which they dye a scarlet colour so strong and fair, that neither the heat of the sun nor the violence of the rain will change it, and the older it is, the better it looks." Of course, the purple-fish (*Purpura lapillus*) is here meant.

Locke also speaks of the "oyster or cockle."

The Latin *cochlea* is properly a snail; but *cochlear* (*cochleare*, or *cochlearium*), "a spoon," or "spoonful," seems to be derived from the form of a bivalve shell, rather than of a snail; it was also a measure for liquids, and in medicine it still signifies a spoonful, hence the Italian *cucchiajo*, French *cuiller*. *Cochlearium* was also used by the Romans for *any* small shell, as in mediæval times. Some authors, indeed, say the spoon was called *cochlear*, not from its shape, but from the pointed end or handle being used for taking the snails (*cochleæ*) out of their shells and eating them, and the broader part for eating eggs, etc. This may be doubted, but a spoon could scarcely resemble a snail-shell, and Martial says (xiv. 121), "Sum cochleis habilis, nec sum minus utilis ovis."

At the meeting of the Ethnological Society, March 4th, 1862, Mr. G. W. Earl gave an interesting description of the singular Malayan shell-mounds, which were formed entirely of *cockle-shells*. He described them as existing in the province of Wellesley, near the Mudah river; that they were about five or six miles from the sea, situated on sandy ridges that appeared formerly to bound the narrow estuaries communicating with the ocean. He adds that these mounds of cockle-shells are about 18 to 20 feet high, and that the Chinese immigrants have largely employed them as a source of lime. These mounds are supposed to be of great antiquity, from the fact of the shells being partly cemented together by crystallized carbonate of lime, the result of the very slow action of atmospheric and aqueous influences. At the bottom of one mound, which contained 20,000 tons of shells, a human pelvis was found; and other remains and stone-implements have been obtained from the Chinese lime-burners. Mr. Earl attributes the formation of these mounds to the Semangs, a diminutive Negro race now sparingly scattered over the surrounding country, but who were evidently very numerous and widely spread in former times.*

In Grey's 'Australia,' vol. i., mention is made of a hill of broken shells, which it must have taken centuries to form, situated between Port George the Fourth, and

* 'Intellectual Observer,' vol. i. p. 239.

Hanover Bay. "It covered nearly half an acre of ground, and in some places was ten feet high; it was situated over a bed of *cockles*, and was evidently formed from the remains of native feasts, as their fire-places and the last small heaps of shells were visible on the summit of the hill." A similar mound noticed near Port Essington, of shells rudely heaped together, is supposed to be a burying-place of the Indians.

At Wigwam Cave, Tierra del Fuego, piles of old shells, often amounting to some tons in weight, were noticed by Dr. Darwin, which had at different periods formed the chief food of the inhabitants.*

These remind us of the so-called kjökkenmöddings (kitchen heaps) of Denmark, or shell mounds, to which the attention of archæologists has been recently attracted in Northern Europe, and which consist of thousands of shells of the oyster, cockle, and other edible mollusks, with implements of stone, such as flint knives, hatchets, etc., and implements of bone, wood, and horn, with fragments of coarse pottery mixed with charcoal and cinders.†

Quite recently, one of these kjökkenmöddings has been discovered at Newhaven, in Sussex, and among the objects found were limpet and other shells, with bones of animals.‡

In 1863, Sir John Lubbock published in the 'Natural History Review' an account he had received from the Rev. G. Gordon, of Scotch kjökkenmöddings on the Elginshire coast, resembling those in Denmark. Mr. Gordon says:—"By far the most striking if not the

* Darwin, ' Voyage of Adventure and Beagle,' vol. iii. p. 234.
† Sir Charles Lyell's ' Antiquity of Man.'
‡ 'Intellectual Observer,' vol. vii. p. 233.

most ancient of the kjökkenmöddings we have in our vicinity is that one which lies within a small wood on the old margin of the Loch of Spynie, and on a sort of promontory formed of those raised shingle beaches so well developed in that quarter. This mound, or rather two mounds (for there is an intervening portion of the ground which has no shells), must have been of considerable extent. A rough measurement gives 80 by 30 yards for the larger, and 26 by 30 for the smaller portion. The most abundant shell is the periwinkle; next in order as to frequency is the oyster, which, as well as those who had it as a large item in their bill of fare, has passed away from our coasts. Save in some of the nooks of our Firth, as at Cromarty, Altirlie, and Avoch, we know not where a small dish of them could be procured. As third in order, in this mound, is the mussel, and then the cockle."

Mr. Gordon further adds that similar refuse-heaps are found all round the shores of the Moray Firth, and that the farmers gradually cart them away to serve as manure or topdressings.

These shell-mounds, Sir John Lubbock states, are actually called "shelly-meddings" by the fishermen of that district.

Sir Gardner Wilkinson found large masses of cockle-shells embedded in the ditches of an old British camp or earthwork, called "Nottle Tor," in the seignory of Gower, in Glamorganshire. This camp stands on a high rock above the sea, and at some distance from any dwelling-house; the shells therefore are from fish eaten by the ancient Britons.

Cockle, mussel, and oyster shells are often discovered in great quantities on the sites of Roman stations.

In the reign of King John, we read of vessels called "cogs." They were supposed to be short and of great breadth, like a *cockle-shell,* whence they are said to have derived their name. The name "*cog*" was variously written, viz. kogge, gogga, kogh, cocka, coqua, etc. "Cogs" were used for the conveyance of passengers from England to France, and as coasting vessels.*

Cockle Soup.—Scald, drain, beard, and wash carefully four dozen of cockles, reserving their liquor in a pan, Put 4 ounces of butter into a stewpan to barely dissolve over the fire; mix in 4 ounces of flour; moisten with a pint and a half of good white stock or milk; season with nutmeg, a pinch of cayenne, and a teaspoonful of anchovy; add half a pint of cream; stir over the fire for a quarter of an hour's gentle boiling, and then, having cut the cockles in halves, pour the hot soup over them in the tureen.†

Cockle Sauce.—Clean cockles thoroughly from all particles of sand, put them into a saucepan with the liquor and a little water, thicken with flour and butter, adding pepper, salt, a little mace, and some cream.

Soyer's Porridge of Cockles, oysters, or mussels, for the poor. They make a most nourishing and palatable food, and on the coast a very economical one.—Take two dozen oysters, or if you use cockles or mussels take a quart of either, put them into an earthen jar with their liquor and three tablespoonfuls of flour; place it on the fire and stir them round and round; add a little salt and pepper and they are done. Eat them thus or add them to soup or porridge. A little dripping or lard is an improvement, also a bay leaf, mint, or an onion sliced.

* See Hist. of the Royal Navy, by Sir N. H. Nicolas, vol. i., note, p. 128.
† Francatelli's 'Cook's Guide.'

Scalloped Cockles.—Wash the cockles well, then scald some dozens of them; strain the liquor into a stewpan, and add thereto 2 ounces of butter, mixed with 2 ounces of flour, a little cream, anchovy, nutmeg, and cayenne; stir the sauce over the fire to boil and reduce for ten minutes, then add a couple of yolks of eggs, and a little lemon-juice, and some chopped parsley; add the cockles; stir altogether over the fire for a few minutes, and fill some scallop shells with this preparation. Cover them over with a thick coating of fried bread-crumbs; place them on a baking-sheet in the oven for five minutes, and serve them quite hot.*

To Stew Cockles.—Clean them and wash them from the sand in three or four waters; boil them and pick them out of the shells. To a pint of the fish put half a pint of fish-stock, 2 ounces of butter, and some pepper and salt; add a spoonful of flour, stirred in gradually, and simmer over a slow fire until it is of a proper thickness; add a large spoonful of essence of anchovy and one of mushroom ketchup.†

The natives of the seignory of Gower cook cockles in various ways; sometimes they fry them with ham. They also make excellent pies of cockles with chopped chives, a layer of bacon being placed at the bottom of the dish; or they fry the cockles with oatmeal and chives, or oatmeal alone; they also make of them an excellent and nutritious soup.

In Ireland, the common cockles are cooked in their shells over the fire, and eaten with oaten cake. The shells are separated by twisting them apart, and a little butter is put into the shell, which is then placed on the turf-fire till the fish inside is fried.

* Francatelli.   † Murray's Modern Domestic Cookery.

Mr. Blackburn, in his 'Travelling in Spain in the Present Day,' says, that one of the best dishes at Seville is composed of rice, pimentoes, cockles (including sand and shells), well boiled in oily gravy.

CARDIUM RUSTICUM or TUBERCULATUM, Linnæus. *Red-nosed Cockle.*—Shell nearly 3 inches in length, and 2 in breadth; very solid, subrotund, opaque, with 21 or more broad ribs which radiate from the beaks, with knots or tubercles on them, which on the anterior slope are flat, and even wanting in young specimens, and on the posterior side are more pointed and rugged; the interstices between the ribs coarsely striated. Umbones prominent; beaks incurved. Ligament large, central tooth large, and the lateral teeth remote.

This large handsome cockle is essentially a Mediterranean species, and is rare and local in England. It is found on the Devonshire coast, at Paignton, and occasionally at Dawlish, and at certain times of the year, especially in the spring after a gale from the east, numbers may be gathered. On paying a visit to the Paignton sands, for the purpose of shell collecting, in the spring of 1862, the beach was quite strewn with broken single valves of this cockle, and there had evidently been quantities of live specimens washed up as well, as we met many persons returning home with their baskets heavily laden with them.

Cardium rusticum varies in colour, from nearly white to a rich-rufous brown; sometimes there is a white band round the shell and one of a dark chestnut-brown towards the margins. The colouring of the animal is most beautiful, the body being of a pink or pale vermilion, the mantle yellow or reddish, and the long foot of a most brilliant crimson. This foot terminates in a hooked point, and when stretched to its utmost is nearly 4 inches in length. It is by means of this organ, that the cockle can bury itself in the sands, and also take those wonderful leaps of which we read in Mr. Gosse's interesting work, 'The Aquarium,' and again in his 'A Year at the Shore,' where he mentions that a specimen was seen to throw itself over the gunwale of a boat when laid on the bottom boards. Mr. Gosse states, in this latter work, that the mode of leaping is performed as follows :—" The long taper foot is thrust to its utmost and feels about for some resisting surface, a stone for instance, which it no sooner feels than the hooked point is pressed stiffly against it, the whole foot, by muscular contraction, is made suddenly rigid, and the entire creature,—mantle, siphons, foot, shell, and all,—is jerked away in an uncouth manner."

There is another cockle found also at Paignton, which is even more scarce than *Cardium rusticum*, viz. *Cardium aculeatum*; it is larger and not so solid, with long spines on each rib, and is of a pale brownish-pink or flesh colour. It is very good to eat. I have had splendid specimens sent to me, alive, from Paignton, in a jar, with seaweed; some measuring more than 3 inches in length and 2½ in breadth, and I have taken them myself at Langston Point, near Dawlish. The foot of the animal is long, and of a reddish-pink, but not nearly so vivid or brilliant in colour as that of *Cardium rusticum*. It is also an inhabitant of the Mediterranean.

Paignton method of Cooking the Red-nosed Cockle.— Cleanse them for a few hours in cold spring water, and then fry them in a batter made of bread-crumbs.*

* Forbes and Hanley, Brit. Moll. vol. ii. p. 15.

Cockle Soup.—After the cockles have been well washed, place them in a stewpan over a slow fire till they open, and then take them out of their shells. Put an ounce of butter or lard, some finely-chopped parsley, a sliced onion, a little pepper, and a teaspoonful of anchovy, into a saucepan, with a little flour, and fry till it becomes brown. To this add a pint of water, or a pint and a half of milk, and, when it boils, place in your cockles. Let it boil again for half an hour, then serve. The cockles, being large, will require to be cut in halves or quarters, previous to their being put into the soup; and the quantity required would be about 2 lb. weight.

In the Bay of Naples, where these cockles abound, they are eaten, as we are told by Poli,[*] either raw or cooked with oil, pepper, salt, herbs, and bread-crumbs, and are called *cocciola* at Naples, and *cappa tonda* at Venice; and Major Byng Hall[†] speaks of cockles stewed in oil as being greatly prized by the natives of Madrid.

Fam. SOLENIDÆ.

SOLEN.—RAZOR-SHELL.

SOLEN SILIQUA, Linnæus. *Razor Shell.*—Shell straight, open at both extremities. Two teeth in left valve, and one in the other; exterior covered with an olivaceous epidermis, concentrically striated. Breadth 1 inch, length from 7 to 8 inches.

The *razor* or spout-fishes are all good for food, but *Solen siliqua,* which is the largest of our British species, is the one generally collected for that purpose. *Solen ensis*

---

[*] 'Testacea utriusque Siciliæ,' 1795.
[†] 'Queen's Messenger,' p. 341.

is eaten in the Feroe Isles, and is there called *langskoel;* and *Solen marginatus,* commonly known as *vagina,* is greatly prized as an article of food by the Neapolitans. This last-named species has a wide range abroad, but is not so common in this country as the two above-mentioned shells, though it is abundant in some localities, amongst others Rye, Tenby, and the Channel Islands.

The razor-shell is the *aulo* of the Romans; and Aristotle, in his 'History of Animals,' gives a description of it, stating that "it buries itself in the sand; can rise and sink in, but does not leave, its hole; is soon alarmed by noise, and buries itself rapidly; and that the valves of the shell are connected together at both sides, and their surface smooth."*

In the time of Athenæus it was much eaten, and highly valued, if we may judge from the following quotations in his 'Deipnosophists:'—

"Araros says, in his 'Campylion'—

"'These now are most undoubted delicacies,
Cockles and solens.'

And Sophron says, in his 'Mimi'—

"'A. What are these long cockles, O my friend,
Which you do think so much of?
B.   Solens, to be sure;
This, too, is the sweet-flesh'd cockle, dainty food,
The dish much loved by widows.'"†

Again, Athenæus says:—"But the *solens,* as they are called by some, though some call them αῦλοι and δόνακες, or pipes, and some, too, call them ὄνυχες, or claws, are very juicy, but the juice is bad, and they are very glutinous. And the male fish are striped, and not

---

* Forbes and Hanley, Brit. Moll. vol. i. p. 240.
† Athenæus, vol. i. b. iii. p. 144, Bohn's Classical Library.

all of one colour, but the female fish is all of one colour, and much sweeter than the male; and they are eaten boiled and fried, but they are best of all when roasted on the coals till their shells open. And the people who collect this sort of oyster are called *solenistæ*, as Phænias the Eresian relates in his book which is entitled 'The Killing of Tyrants by way of Punishment;' where he speaks as follows:—' Philoxenus, who was called the Solenist, became a tyrant from having been a demagogue. In the beginning he got his living by being a fisherman, and a hunter after *solens* ; and so, having made a little money, he advanced and got a good property.' "

On some parts of our shores great quantities of razor-shells are collected, sometimes by putting a little salt on the holes, which irritates the fish and makes it rise to the surface; and again in the following manner, as described by Messrs. Forbes and Hanley :—" A long narrow wire, bent and sharpened at the end, is suddenly thrust into the hollows of the sands indicative of the presence of these animals, and, passing between the valves, the barbed portion fixes itself, on retraction, in the animal, and forces it to the surface." At Tenby baskets-full are often brought to the door, and they are considered very good to eat. In Japan they are said to be so highly prized that, by the express order of the prince of that country, " it is forbid to fish them, until a sufficient quantity hath been provided for the emperor's table."*

In the Bay of Concepcion are several species of shell-fish highly esteemed, and Ulloa especially mentions some *Venuses* and a number of razor-shells.

* ' Glimpses of Ocean Life,' by John Harper, F.R.S.

The German name for this shell is *scheidenmuschel* or *messerschalenmuschel*, and the French call it *manche de couteau*, and *coutoye*, and the Andalusians, *muergo*.

At Naples it is considered a great delicacy, and quite a *recherché* morsel, too expensive for the common people, a dishful selling at 6 carlines, which is equal to 2s. of our English money. It is cooked in the following manner:—

Razor-fish Soup.—Take 2 lb. of razor-fish, and, after they have been well washed, put them into a saucepan, and keep them on a slow fire till they open, then take out the fish from the shells. Chop up some parsley very fine, and put it, with a tablespoonful of oil or an ounce of butter, into a saucepan, and fry until it becomes brown. To this add a pint of water, or a pint and a half of milk, and, when boiling, place in your fish, with a little salt and pepper, and let it boil again for half an hour. Add toasted bread before it is served up, or boil some vermicelli with it, of course adding more water.

To cook Razor-fish.—Boil them for ten minutes or so, then take them out of their shells, and fry them with butter or lard. Add a little salt and pepper.

Another way to cook "Solens."—Stew them in milk till they are tender, add pepper, salt, and butter is a great addition.

The razor-fish is much prized on the Scotch coast, where it is merely boiled, and eaten with salt and pepper. Poli says that it is good either raw, or fried with breadcrumbs, pepper, oil, and lemon-juice.

Fam. CYPRINIDÆ.
ISOCARDIA.—OXHORN-COCKLE.

ISOCARDIA COR, Linnæus. *Heart-shell or Oxhorn Cockle.*—Shell very strong, nearly spherical, heart-shaped, concentrically striated, equivalve, smooth, with a dark reddish-brown epidermis; beaks very prominent and curled; two primary teeth in the right valve, lying parallel to each other; in the left valve, the outer tooth is indented and is large, the other thin and laminar. The lateral tooth strong and elongated, situated under the ligament, which is external.

This magnificent mollusk is very partially distributed, though plentiful in some places. Specimens have been sent to me from Dublin Bay, where I grieve to say they are getting very scarce, and also from Brixham, where they are highly prized by the fishermen. They do not however often bring them on shore, though they bring them up in the dredges, unless they wish to make a present of a dish to some friend, or know where they can dispose of them. They call them "Torbay-noses," and they are also known by the names of "Oxhorn-cockles" and "Heart-shells;" in France, *cœur de bœuf*, in Holland, *zots-kappen*, or fool's cap, and at Naples, *cocciola zizza*. Mr. J. G. Jeffreys, quoting an interesting account of *Isocardia cor*, by the Rev. James Bulwer (who kept a specimen in a vessel of sea-water, and was therefore able to study the habits of the animal), given in the 'Zoological Journal,' states, "that the animal appears insensible both to sound and light, as the presence or absence of either did not interrupt its movements, but its sense of feeling appeared to be very delicate; minute substances being dropped into the orifice of the mantle instantly excited the animal, and a column of water strongly directed, expelled them from the shell. With so much strength was the water in some instances ejected that it rose above the surface of three inches of superincumbent fluid. . . . Locomotion very confined; it is capable, with the assistance of its foot, which it uses in the same manner (but in a much more limited degree) as the *Cardiacea*, of fixing itself firmly in the sand, generally choosing to have the umbones covered by it, and the orifices of the tubes of the mantle nearly perpendicular."*

"Resting in this position on the margin of a sand bank of which the surrounding soil is mud, at too great a depth to be disturbed by storms, the *Isocardia* of our Irish Sea patiently collects its food from the surrounding element, assisted in its choice by the current it is capable of creating by the alternate opening and closing of its valves."

The Mediterranean species of this bivalve are smaller than those found on our coasts, and there are no less than "five or six kinds known in the European and Indian seas."†

Epicharmus, in his play of the 'Marriage of Hebe,' mentions shell-fish of all kinds, and says:—

"And bring too the black
Cockle, which keeps the cockle-hunter on the stretch."‡

This may possibly refer to the oxhorn-cockle.

The wife of a coastguardsman who had lived many years at Brixham, and had often luxuriated in a dish of

---

* Brit. Conchol., by Jeffreys, vol. ii. pp. 300, 301.

† 'Manuel de Conchyliologie,' par le Dr. J. C. Chenu.

‡ Athenæus, Bohn's Class. Lib. b. iii. p. 142.

these delicious shell-fish, gave me the following recipe for cooking them :—

To dress Torbay-noses.—" Wash the shells, then boil them for ten minutes or so; take the fish out of the shells and put them into a frying-pan with some butter, a little salt and pepper, and fry till they are of a good brown colour; then serve."

Fam. MYTILIDÆ.

MYTILUS.—MUSSEL.

MYTILUS EDULIS, Linnæus. *Common Mussel.*—Shell equivalved, wedge-shaped, rather pointed at the beaks. In the hinge are three or four tooth-like crenulations. Ligament internal or nearly so, and very strong. Colour of the shell a greyish blue, sometimes radiated with darker blue. Epidermis olivaceous.

The mussel is called in Anglo-Saxon *muscl, muscel, muscule, muscla,* which names mean that which instantly retires on being touched; in Dutch, *mossel,* in Danish, *muskel,* in German, *muschel,* in French, *moule,* at Bordeaux, *charron* (from the village of that name, where there is a large mussel trade) ; in Feroese, *kreaklingur,* and in Andalusia, *longherone.* Mussels are used for food in many places, and also for bait, "and on some parts of the Northumberland coast the fishermen have made *mussel gardens* for the preservation of these shell-fish ; they are formed by piling up stones round certain places on the seashore, between tide-marks, and are carefully watched by their proprietors."*

M. de Quatrefages, in his interesting work, ' Rambles

* 'A Book for the Seaside,' p. 100.

of a Naturalist,' gives an account of the origin and development of the mussel-trade on the French coast. "An Irishman of the name of Walton was shipwrecked on the coast in 1235, near the little village of Esnandes, in the Bay of Aiguillon, and was the only person saved out of all the crew of the ill-fated vessel. He amply repaid the services which had been rendered him; some sheep were saved from the wreck, which he crossed with the animals of the country, producing a breed of sheep which is still held in high estimation. He invented a kind of net, the 'allouret,' for catching shore birds which skim the surface of the water at twilight or dark, and in order to make these nets thoroughly effective, it was necessary to go to the centre of the immense bed of mud, where the birds sought their food, and to secure a number of poles to support the nets, which were between 300 and 400 yards in length. On examining these poles, Walton discovered that they were covered with *mussel spawn*. He then increased the number of his poles, and after various attempts he constructed his first artificial mussel-bed, or *bouchot*. At the level of the lowest tides he drove into the mud stakes that were strong enough to resist the force of the waves, and placed them in two rows about a yard distant from each other. This double line of poles formed an angle, whose base was directed towards the shore, and whose apex pointed to the sea. This palisade was roughly fenced in with long branches, and a narrow opening having been left at the extremity of the angle, wicker-work cases were arranged in such a manner as to stop any fishes that were being carried back by the retreating tide. It was soon found inexpedient to trust only to the chance of the currents and waves that might bring in the young mussels to the poles and fences, and men frequently went to a very great distance in search of the young mollusks,—even as far as the plateau of Chatelaillon."

M. de Quatrefages further tells us that the little mussels that appear in the spring are called *seeds;* they are scarcely larger than lentils till towards the end of May, when they rapidly increase in size, and are then called *renouvelains,* and in July are ready for transplanting. They are detached from the *bouchots* which are situated at lowest tide-mark, and are then put into pockets or bags made of old nets, " which are placed upon the fences that are not quite so far advanced into sea." The young mussels attach themselves by means of their *byssus* all round the pockets or bags. As they increase in size and become crowded together, they are taken out and distributed over other poles lying nearer the shore, and the full-grown mussels which are ready for sale are planted on the *bouchots nearest* the shore. The fishermen gather enormous quantities of fresh mussels every day, and take them in carts or on the backs of horses "to La Rochelle and other places, from whence they are sent as far as Tours, Limoges, Bordeaux."

"It appears that the French mussel-breeders have discovered that mussels which live suspended to piles, or ropes of vessels, nets, etc., attain to a larger size, than those which live on the bottom, be it sandy, rocky, or muddy; they therefore suspend thick ropes to wooden piles, and the mussels adhere by their byssus to them, the ropes are then tightened a little to prevent the animals lying on the bottom."*

The Billingsgate market is chiefly supplied with mus-

* Phipson's 'Utilization of Minute Life,' pp. 163, 164.

sels from Holland, the east coast of England, Cornwall, and Devonshire, in August and September, though smaller quantities are received from other parts of our coasts, besides those above mentioned. About ten or twenty tons weight arrive at a time, though, of course, the quantity varies according to the season, and they are sold at 1*s*. a measure. In the evidence given before the Fisheries Commission, at Exeter, December 24, 1863, it was stated, that the price of these shell-fish taken in the estuary at Lympstone was 8*s*. per sack of ten pecks, but that the supply was decreasing.

Dr. Knapp informed Messrs. Forbes and Hanley that the quantity of mussels consumed in Edinburgh and Leith is about 10 bushels per week, "say for forty weeks in the year, in all 400 bushels annually. Each bushel of mussels, when freed and shelled from all refuse, will probably contain from 3 to 4 pints of the animals, or about 900 to 1000, according to their size. Taking the latter number, there will be consumed, in Edinburgh and Leith, about 400,000 mussels. This is a mere trifle compared to the enormous number used as bait for all sorts of fish, especially haddocks, cod, ling, halibut, plaice, skate, etc.; and at Newhaven, the total consumption of mussels for bait may be reckoned at 4,320,000 annually. There are nearly as many used at Musselburgh, Fisherrow, etc., and other places on the Frith of Forth, and we may calculate that 30,000,000 or 40,000,000 of mussels are used for bait alone by the fishermen of that district each year."*

The mussel has the power of attaching itself by means of its "byssus" to rocks and stones; and we read that the bridge at Bideford, in Devonshire, cannot be kept

* Forbes and Hanley, Brit. Mollusca, vol. ii. pp. 174, 175.

in repair by mortar, owing to the rapidity of the tide. "The corporation, therefore, keep boats to bring mussels to it, and the interstices of the bridge are filled by hand with these shellfish, and it is supported entirely by the strong byssus or threads these mussels fix to the stonework."*

This byssus proceeds from a gristly shaft, which, Mr. Jeffreys states, appears to support the bundle of filaments like the handle of a broom; and Aristotle mentions this shell-fish in his list of cartilaginous fish.

So valuable are mussels towards the protection of the shores from the inundations of the sea on some parts of our coasts, that it becomes necessary to prevent their being gathered in some places (see 'Times,' August 7th, 1865). An action for trespass was brought some time ago for the purpose of establishing the right of the lord of the manor to prevent the inhabitants of Heacham from taking mussels from the seashore. The locality is the foreshore of the sea, running from Lynn in a north-westerly direction towards Hunstanton, Norfolk; and "the nature of the shore is such that it requires constant attention, and no little expenditure of money, to maintain its integrity, and guard against the serious danger of inundations of the sea." A large quantity of shingle, seaweed, and mussels is always to be seen, and beds of mussels extend for miles along the shore, and mix with the seaweed and shingle, which get fixed on the artificial jetties running into the sea, attaching themselves by means of the byssus to these embanking defences, thereby rendering them firm, and thus acting as barriers against the sea; therefore, while it is important for the inhabitants, who claim a right by custom, to

* 'Glimpses of Ocean Life,' p. 179.

take mussels and other shellfish from the shore, it is equally important for the lord of the manor to do his utmost to prevent these natural friends of his embankments and jetties from being removed in large quantities from his part of the shore.

Neumann tells us that calcined mussel-shells make strong lime, and bind quickly, and that shell-lime is generally considered stronger than stone-lime. Mussel-shells, when polished, make pretty pincushions and needlebooks, and at the colourists, they are filled with gold, silver, and bronze, and sold for heraldic painting and illuminating. It was in one of these shells, also, in which the witch, in the quaint old story, put to sea for the purpose of wrecking her enemy's ships.

A large species of mussel, called *awabi* or *awabee*, is used in Japan as a new year's gift. The day is spent in paying respects, visiting, and giving presents to friends and relatives, and they mostly consist of *awabi*. *Awabi*, in days of yore, were the first sustenance and support of the Japanese, as acorns were formerly the primitive diet of the inhabitants of Europe, and the *awabi* is the emblem, or rather the memorial, of the frugality of their forefathers.*

There is another purpose for which these shells are used, which would astonish the "Truefitts" of the present day; for Grey, in his 'Australia,' mentions that amongst the contents of a native woman's bag was a mussel-shell for cutting the hair.

There is an interesting account in Captain O'Brien's 'Adventures during the late War,' of the method of fishing for mussels in the Bay of Concepcion. A man and woman in a canoe push off from the shore, to a certain

* 'Religious Ceremonies,' vol. iv. p. 315.

depth, when the man with a long pole ascertains the depth of the mussel-bed. This pole, which has a sharpened end, is struck into the bed, and serves as the anchor or mooring for the boat; the woman, with her arms round it, makes it her line of descent. With this, as a conductor, she slides or slips down, and soon reappears, with her arms crossed round the pole, but with both hands as full as they can hold of mussels. Having deposited her handfuls in the canoe, she descends again and again six or eight times, until her cargo is complete. Upon Captain O'Brien's remonstrating with a man for imposing such a dangerous duty upon a woman, instead of undergoing it himself, he explained to him, that this diving was a privilege of the sex, and that no man would dare to be so unmanly as to rob a woman of her birthright. These Chilian, or Bay of Concepcion belles, sell their produce in the market for dresses and finery.

The usual size of the common mussel is about two inches and a half in length, and about half that breadth; but in 1862 I procured two specimens from Exmouth, which had been dredged, the largest measuring five inches in length and two and a half in breadth, the other four inches long and one and a quarter wide. Mr. Jeffreys also mentions having a specimen which measures nearly five inches in length. Though mussels are a valuable article of food, and considered wholesome, yet many cases of poisoning by mussels have occurred ; but it may generally be traced to their having been gathered from either the sides of docks, or piers, where there are copper bolts or nails, or from ships that are copper-bottomed ; or else from the neighbourhood of large town sewers, the sewerage water running over the rocks on which the mussels grow. In the 'Field,' November

15th, 1862, is an interesting account of an experiment made on oysters that had become so impregnated with copper as to be as green as verdigris. They were taken from Falmouth harbour. An attempt was made to extract the copper from them; and, after putting a hundred or more into a large crucible, reducing them to ashes, and continuing to increase the heat until the copper was melted, the produce was a bright bead of pure copper, which, according to the description, would be about the size of a large pin's head. Mr. Penwarne, who communicated this article to the 'Field,' adds, that the oysters may have lain on a lode, or the copper might have accumulated from the wash of the stamping-mills. This proves, without doubt, that shell-fish can be impregnated with copper, or other poisonous substances, which probably would affect those who ate them. Some persons consider that mussels are unwholesome if a small species of crab (*Pinnotheres pisum* or *Pinnotheres veterum*), which is sometimes found in their shells, is not carefully taken out; others, that they are only fit for food in the winter months; and by some on account of their feeding on the spawn of the star-fish, which is poisonous.* It is said that if a silver spoon is boiled with the mussels, and it turns black, it proves that they are poisonous, and not fit to be eaten. But, whatever may be the cause of the wholesale poisoning by these shell-fish, they have been the means of saving many poor from starvation in times of scarcity. Mr. Patterson, of Belfast, in his 'Introduction to Zoology,' mentions having been informed by an old inhabitant of Holywood, near the above-mentioned town, that in 1792 or 1793 there was a great drought

* Jeffreys, Brit. Conchology, vol. ii. p. 109.

prevailing, which caused much distress, and that, in the month of June or July, twenty poor families from the interior of the country encamped on the roadside, near the beach to the west of Holywood, remaining there about five weeks, subsisting partly on such vegetable matter as they could pick up about the hedgerows and fences, but principally upon the mussels which are so abundant on the extensive mud-banks of the neighbouring coast. No instance of disease from this diet occurred, and during that summer the poorer classes in the village appeared quite as healthy as in other years, though mussels formed their chief food.

Athenæus says that mussels are moderately nutritious, and digestible; the best being the Ephesian kind, which are particularly good when taken about the end of autumn (vol. i. p. 150).

In the Feroe Isles, the large horse-mussel, *Mytilus modiolus*, is eaten, and they call it in Feroese *ova*. Mr. Alder tells us that at Rothesay they are collected for food* (though not so delicate as *Mytilus edulis*), and in the Shetland Isles for bait, where they are known by the name of *yoags*. They are also eaten in the north of Ireland, but not considered very good, on account of their strong scent and flavour; but they are capital bait for cod. At Tenby they call *Mytilus modiolus* the poisonous mussel, and affirm that no one ever ventures to eat it.

Pearls are occasionally found in the common mussel, and also in the oyster, scallop, cockle, periwinkle, and pinna; but they are generally inferior in size and quality to those of the freshwater pearl-mussel, *Unio margaritiferus*; and Mr. Beckman, in his 'History of

* Forbes and Hanley, Brit. Mollusca, vol. ii. p. 185.

Inventions,' states that real pearls are found under the shield of the sea-hare (*Aplysia*), as has been observed by Bohadsch, in his book ' De Animalibus Marinis' (Dresdæ, 1761). Our Scotch pearl-fishery has, within the last few years, been most successfully revived; and in 1860 Mr. Moritz Unger, a foreigner, on making a tour through the districts where the pearl-mussel abounds, found that the pearl-fishing was not altogether forgotten, many of the people having pearls in their possession, of which they did not know the value. He purchased all he could obtain; consequently, in the following year, many persons devoted their spare time to pearl-fishing, and during the summer months made as much as £8 to £10 weekly. The summer of 1862 was most favourable for fishing, owing to the dryness of the season, and the average price was from £2. 6s. to 10s., £5 being a high price. They now fetch prices varying from £5 to £20. The Queen purchased one Scotch pearl for 40 guineas; others at high prices have been bought by the Empress of the French and the Duchess of Hamilton; and Mr. Unger has a necklace of these pearls valued at £350.* Pearl-mussels are found in Lochs Earn, Tay, Rannoch, and Lubnaig, and in the Don, the Leith, and in many of the other Scotch streams; also in some of the Welsh rivers, from whence I have received fine specimens; in Ireland, near Enniskillen, and in the river Bann, which is noted for its fine pearls. They wade for them in the shallow pools, or take them by thrusting a long stick between the valves when the shell is open. When a number have been collected they are left to decompose, when the pearls drop out.† They may also be

---

* The 'Times,' December 24th, 1863.   † Tour in Ulster.

found in Kerry, in Donegal, in the Moy near Foxford, and in many of the other Irish rivers; and Mr. Buckland states, in the 'Field,' December 10th, 1864, that they abound near Oughterard, and that a man called "Jemmy the Pearl-catcher" told him he knew when a mussel had a pearl in it, without requiring to open it first, because "she (the mussel) sits upright with her mouth in the mud, and her back is crooked,"—that is, it is corrugated like a cow's horn. Bruce, in his "Travels," observes that the pearl-fishers of Bahrein informed him that they had no expectation of finding a pearl when the shell was smooth and perfect, but were sure to find some when the shell was distorted, and deformed; and he adds that this applies equally to the Scotch pearl-mussels. In France they also collect pearls from the pearl-mussels, and they generally sell them as *foreign* pearls. At Omagh, in the north of Ireland, there was formerly a pearl-fishery; and Gilbert, Bishop of Limerick, about 1094, sent a present of Irish pearls to Anselm, Archbishop of Canterbury. Scotch pearls were in demand abroad as early as the twelfth century. Suetonius says that the great motive of Cæsar's coming to Britain was to obtain its pearls, and states that they were so large that he used to try the weight of them by his hand, and dedicated a breastplate made of them to Venus Genetrix.* According to Pliny, the island of Taprobane (Ceylon) was most productive of pearls; and he considers that the most valuable were those found in the vicinity of Arabia, in the Persian Gulf.

Oriental pearls are found in the *Meleagrina margaritifera*, or pearl-oyster; and Chares of Mytilene, in his seventh book of his 'Histories of Alexander,' tells us

* Camden's 'Britannia,' p. 962.

that "in the Indian Sea, and also off the coast of Armenia, Persia, Susiana, and Babylonia, a fish is caught very like an oyster; large and of oblong shape, containing within its shell flesh which is plentiful, white, and very fragrant, and from it the men pick out white bones, called by them pearls. And of these they make necklaces and chains for the hands and feet, of which the Persians are very fond, as are the Medes and all Asiatics, esteeming them as much more valuable than golden ornaments."* Occasionally, they are called *stones* and *bones* by Greek authors; and Tertullian calls them maladies of shell-fish and warts—"concharum vitia et verrueas." Pliny states† that when pearls grow old they become thick and adhere to the shell, from which they can only be separated by a file; again, that pearls which have one surface flat and the other spherical, opposite to the plane side, are for that reason called *tympania*, or tambour-pearls, "quibus una tantum est facies, et ab ea rotunditas, aversis planities, ob id tympania nominantur." The "tympana," or hand-drums of the ancients, were often of a semi-globular shape, like the kettle-drums of the present day. Shells which had pearls still adhering to them were used as boxes for unguents.‡ Long pear-shaped pearls, called *elenchi*, had their peculiar value, resembling the alabaster boxes in form which were used for ointments. Earrings were invented by the Roman ladies, called "*crotalia*, or castanet pendants, from the pearls rattling as they knocked against each other."§

The story of Cleopatra swallowing the pearl, in order

* Athenæus, vol. i. p. 155.
† Pliny, Nat. Hist. vol. ii. b. ix. p. 432.
‡ Pliny, Nat. Hist. b. ix. p. 432, and *note*.
§ Pliny's Nat. Hist. vol. ii. b. ix. p. 435.

that she might say she had expended on a single entertainment ten millions of sesterces, is too well known to require repeating here; suffice it to say that Pliny informs us, that before the time of Antony and Cleopatra, Clodius, the son of the tragic actor Æsopus, had done the same at Rome; "he, having dissolved in vinegar (or at least attempted to do so) a pearl worth about £8000, which he took from the earring of Cæcilia Metella."* Pliny further adds that, by way of glorification to his palate, Clodius Æsopus was desirous of trying what was the taste of pearls, and as he found it wonderfully pleasing, that he might not be the only one to know it, he had a pearl set before each of his guests for him to swallow."† It was not unusual for the Romans to adorn their horses, and other favourite animals, with splendid necklaces;‡ and we are told that Incitatus, the favourite horse of the Emperor Caligula, wore a pearl collar. The Roman ladies even wore pearls at night, that in their sleep they might be conscious of the possession of these valuable gems. Julius Cæsar prohibited the use of purple and pearls to all persons who were not of a certain rank, and the latter also to unmarried women.§ From the twelfth to the sixteenth centuries, extravagance in jewellery was carried to an unlimited extent at the Courts in Europe; and from the reign of Francis I. to that of Louis XIII., the greater part of the jewels worn were set with pearls, and these latter were worn in preference to all other ornaments until the death of Maria Theresa of Austria.‖ The French call irregular-shaped pearls "Perles barroques,"

---

* Hor. ii. Sat. iii. 239.   † Pliny, Nat. Hist. vol. ii. b. ix. ch. 59.
‡ Smith's Dict. of Greek and Roman Antiquities: "Monile," p. 768.
§ 'Gems and Jewels,' p. 27, Madame de Barrera.
‖ 'Gems and Jewels,' p. 58.

and these malformations were ingeniously utilized by the fanciful taste of the Cinque Cento period.* No doubt many of my readers will remember the specimens exhibited in the loan collection at the South Kensington Museum. One was a Cinque Cento pendant in the form of a siren; the head, neck, and arms of white enamel, the body made of a very large "pearl barroque," and a fish's tail enamelled, and set with rubies. It belonged to Colonel Guthrie, and is of fine Italian work of the sixteenth century. Another, in the possession of Messrs. Farrer, was a gold pendant jewel in the form of a ship with three masts, a large pearl barroque forming the hull, etc. The wedding dress of Anne of Cleves was "a gown of rich cloth of gold, embroidered with great flowers of large orient pearls." The unfortunate Mary, Queen of Scots, possessed pearls which were considered the finest in Europe, and these were purchased in a most iniquitous manner by Queen Elizabeth from the Earl of Moray, for a third part of their value. Miss Strickland states (in her 'Lives of the Queens of Scotland,' pages 82 and 83, vol. vi.) that if anything further than the letters of Drury and Throckmorton be required to prove the confederacy between the English Government and the Earl of Moray, it will only be necessary to expose the disgraceful fact of the traffic for Queen Mary's costly parure of pearls, her own personal property, which she had brought from France. A few days before she effected her escape from Lochleven Castle, the Regent sent these with a choice selection of her jewels, very secretly, to London, by his trusty agent, Sir Nicholas Elphinstone, who undertook to negotiate their sale with the assistance of Throckmorton. Queen Elizabeth had

---

* 'Precious Stones, Gems,' etc., by Rev. C. W. King, p. 232.

the first offer of them, and the French Ambassador thus describes them:—" There are six cordons of large pearls strung as paternosters; but there are five-and-twenty separate from the rest, much finer and larger than those which are strung. These are, for the most part, like black muscades" (a very rare and valuable variety of pearl, with the deep purple colour and bloom of the muscatel grape).*

They were appraised by various merchants, but Queen Elizabeth was determined to have them at the sum named by the jeweller, though he would have made his profit by selling them again. Others valued them at three thousand pounds sterling; some Italian merchants at twelve thousand crowns; and a Genevese at sixteen thousand crowns; but twelve thousand was the price Queen Elizabeth was allowed to have them for, and Catherine de Medicis was quite as eager to purchase these pearls as her good cousin of England, knowing they were worth nearly double the sum at which they had been valued in London, having presented some of them herself to Mary. She therefore used every endeavour to recover them, but the French Ambassador wrote to inform her that it was impossible to accomplish her desire of obtaining the Queen of Scots' pearls; "for, as he had told her from the first, they were intended for the gratification of the Queen of England, who had been allowed to purchase them at her own price, and they were now in her hands." The possession of wealth and jewels is not always a source of happiness or benefit to their possessors, if we may judge "from the above-mentioned fact in history, and indeed it is even more clearly exemplified in the case of the eminent Mogul,

* See *note*, 'Lives of the Queens of Scotland,' vol. ii. p. 83.

who died of hunger during a grievous famine, which depopulated part of Guzerat. A large mausoleum or Mahometan tomb was erected to his memory in the suburbs of Cambay, with an inscription, telling us that during this terrible scarcity the deceased had offered a measure of pearls for an equal quantity of grain, but not being able to procure it, he died of hunger.*

A pearl is described by Madame de Barrera, as nearly the size of a pigeon's egg, and pear-shaped; it weighed 250 carats, and was known as "La Peregrina," and belonged to the crown of Spain. It was brought from Panama in 1560 by Don Diego de Temes, who presented it to Philip II. "It was then valued at fourteen thousand ducats, but Freco, the king's jeweller, having seen it, said it might be worth £14,000, £30,000, £50,000, £100,000, as such a pearl was priceless." In 1779 a pearl, which from its shape was called the Sleeping Lion, was offered for sale at St. Petersburg, by a Dutchman; it weighed 578 carats, and was bought in India for £4500.

The largest pearl known, I believe, is the one which was exhibited at the South Kensington Museum, in the loan collection, in the possession of A. J. B. Beresford Hope, Esq. It weighs 3 oz., is 2 inches long, 4½ inches in circumference, and is set as a pendant.

The most productive pearl-fishery banks lie on the west coast of Ceylon, between the eighth and ninth degrees of north latitude, near the level dreary beach of Condatchy, Aripo, and Manaar.† The other principal fisheries are those of the Bahrein Islands in the Persian Gulf, Coromandel, Catifa in Arabia (which produced

* Forbes's 'Oriental Memoirs,' vol. ii. p. 18.
† 'Voyage of the Novara,' vol. i. pp. 379, 380, 381.

the pearls purchased by Tavernier for £110,000), the Algerine coast, the Sooloo Islands, and, in the western world, the Bay of Panama and the coast of Columbia, which had formerly some very valuable pearl fisheries, for Seville alone is said to have imported thence upwards of 697 lb. in the year 1587. The hardships and sufferings endured by the divers are very great. After a long dive, we are told that the natives of the Paamuto Islands may be seen squatting on the reefs with blood gushing from the ears and nose, and become quite blind, for ten or twenty minutes.

At the Bahrein fisheries, the trade is in the hands of the merchants, who bear hard on the divers; and even those who make the greatest exertions in diving can scarcely obtain a sufficiency of food.* In Ceylon, the fourth part of the pearls brought up is the diver's share. In each boat are ten divers, each with an assistant. Before the divers descend a number of quaint ceremonies are gone through with incantations, both in the boats and on the shore. So superstitious are these men, that not one of their number, Christian or idolater, would continue their employment without the countenance of the sorcerer, and in 1857 Government was compelled to pay these impostors. The chief sharkcharmer was a Roman Catholic.† The same authority further states that the utmost depth in which a diver can remain safely is about seventy feet. They can remain under water from fifty to sixty seconds, and the diving is carried on from five to six hours daily; and each of the ten divers can in the course of the day bring up from 1000 to 4000 pearl shells. A single oyster contains some-

* M'Culloch's Commercial Dictionary.
† 'Voyage of the Novara,' vol. i. p. 332.

times thirty or forty pearls, of which some may be worth a sovereign on the spot. The small valueless seed-pearls are burnt, and sold as pearl lime to the wealthy Malays, to add to the betel and cabbage-nuts which they chew. The Ceylonese mix the lustreless pearls with grain, and feed their poultry with them, in whose crops the pearls regain their former brilliancy after a few minutes' grinding. The crops are slit up, and the pearls are taken out. It is said to be done by other Indian races, but that the pearls lose weight.† In India the priests of Buddha keep up the strange belief as to the origin of pearls which I have mentioned elsewhere, and make it a pretext for exacting what they term "charity oysters" from the divers and boatmen of their faith for the use of Buddha, who, when thus propitiated, will make the fish yield more pearls in future seasons.‡

The pearl fishery of Ceylon in 1864 suffered considerably, owing to an irruption of the skate fish, which was said to have killed the oysters; and the loss of revenue was calculated at £50,000.

The common freshwater *Unio* (*Unio tumidus*), and also *Unio pictorum*, both produce pearls, but they are generally small, and of a bad colour; sometimes I have found several in one shell, and, again, I have opened many, and not been successful.

A species of freshwater mussel, *Anodonta cygnea*, is said to be eaten in the county of Leitrim by the peasantry, and *Unionidæ* are eaten in the south of Europe, either roasted in their shells and drenched with oil, or covered with breadcrumbs and scalloped; and, according to Dr. Wilhelm Gottlob Rosenhauer, *Unio Requienii*

* 'Voyage of the Novara,' vol. i. p. 385.
† 'Household Words:' " My Pearl-fishing Expedition," vol. iii. p. 80.

and *Unio litoralis,* which are found near Granada, in the river Jenil, are often brought to the market; but when the fish are taken out of their shells and cooked, they are described as very tough food. *Anodontæ* and *Unionidæ* (*Anodontes et Mulettes*) are employed by the fishermen in the neighbourhood of Nantes for bait;* and I have occasionally used *Dreissena polymorpha* for the same purpose, which seemed to be greatly appreciated by the fish in the pond where I was fishing, as they greedily sucked off the bait as fast as it was put on the hook. The *Dreissenæ* were brought from the canal at Sawley, Leicestershire, and turned into the ponds, where they have thriven wonderfully, and are the favourite food of water rats, if one may judge from the number of empty shells deposited on the banks amongst the rushes, in small heaps sometimes two or three inches deep. In some countries the shells of the large *Anodontæ* are used for skimming milk.

In the north-western part of Australia, a freshwater mussel forms a staple article of food, while in the southwestern part of the continent the natives will not touch them, but regard them with a superstitious dread and abhorrence. In Grey's 'Australia,' he gives an account of a native, Kaiber by name, whom he ordered to gather some of these shellfish for food, as they were nearly dying from hunger, but the man steadfastly refused, as he affirmed that by touching them, the native sorcerer, or Boyl-yas, would acquire a mysterious influence over him, which would end in his death. At last, however, he was ordered to bring some instantly, as Mr. Grey intended eating them. After thinking for a moment or so, Kaiber walked away for this purpose, but bitterly

* 'Catalogue des Radiaires,' etc., par Frédéric Cailliaud de Nantes.

lamented his fate whilst occupied with his task. It was true, he said, he had not died of hunger or thirst, but this was all owing to his courage and strong sinews; yet, what would these avail against the supernatural powers of the Boyl-yas. " They will eat me at night, whilst worn-out by fatigue I must sleep." However, the mussels were brought, and Mr. Grey made a meal of them.* It is not only of late years that *Mytilus edulis* has been thought worthy to grace our tables, for in 1390 we have the following recipes given in a "role" of ancient English cookery, compiled by the master cooks of King Richard II., called the ' Forme of Cury :'—

"*Muskels in brewet (broth)*, 122.—Take muskels (mussels), pyke them, seeth hem with the own broth (in their own liquor). Make a lyor (mixture) of crustes *i.e.* of brede) and vinegar; do in onyons mynced, and cast the muskels thereto, and seeth it, and do thereto powder, with a lytel salt and safron. The samewise make of oysters.

"*Cawdel of Muskels*, 124.—Take and seeth muskels, pyke (pick) hem clene, and waishe hem clene in wyne. Take almandes and bray hem. Take some of the muskels and grynde hem, and some hewe small. Draw (mix up) the muskels yground (that are ground) with the self (same) broth. Wryng the almandes with faire (clean) water. Do all this togider. Do thereto verjous (verjuice) and vinegar. Take whyte of lekes, and parboil hem wel. Wryng out the water, and hew hem small. Cast oile thereto, with onyons parboiled, and mynced small. Do thereto powder, fort, safron, and salt; a lytel seeth it, not to stondyng (too thick), and messe it forth."†

"*Soyer's Recipe for Cooking Mussels.*—Take 3 dozen

* Vol. ii. pp. 84, 85.
† ' Antiquitates Culinariæ,' by the Rev. Richard Warner, p. 23.

mussels, wash them and place in a stewpan over the fire for ten minutes, to open the shell (sometimes a small crab will be found in them, which remove, as they are rather unwholesome); replace them, with their liquor, and bottom shell, in the pan; add a spoonful of flour, mixed with some butter or lard, and a spoonful of chopped parsley; stir it, and stew for five minutes, and serve. If required in large quantities, take the large boiler, put therein 4 lb. of lard or butter, and 4 lb. of sliced onions; fry for five minutes. Have ready 2 pailfuls of mussels out of the shell, and in their liquor, which put in the boiler with 1 lb. of salt, 2 oz. of pepper, 2 oz. of sugar, 2 lb. of chopped parsley, and 2 lb. of flour, mixed with water to the thickness of good cream; boil ten or fifteen minutes, stir it gently with a wooden spatula, and serve. If not required maigre, use instead of water the same quantity of boiling stock mixed with flour; a flavour of herbs may be given if liked, and bits of meat added to it."

"*Mussel Sauce.*—Cleanse, beard, wash, and blanch or parboil two quarts of mussels, take all the white fat mussels out of their shells, and place in a bain-marie, reserving their liquor in a basin. Then knead 4 oz. of butter with 2 oz. of flour, some nutmeg, pepper, and salt, add the liquor from the mussels, a piece of glaze, and half a pint of cream; stir the whole on the stove fire till it boils, and keep it boiling for ten minutes, then add a season of 4 yolks of eggs, and pass through a tammy on the mussels; just before sending the sauce to table, throw in a tablespoonful of chopped and blanched parsley, and a little lemon-juice. This sauce is well adapted for boiled whitings, turbot, cod, haddock, and gurnet."*

* Francatelli's 'Modern Cook.'

"*To Dress Mussels.*—After having well washed and scraped their shells, drain them, and put them to dry into a stewpan upon a good fire, letting them remain till the heat opens them. Then take them out of the shells one by one, being careful to pick off the beards where you find any; and put them into a stewpan, with a bit of butter, parsley, and scallions, shred small; shake them over the fire, and put a little flour, moistening them with broth; when the sauce is consumed, put in the yolks of 3 eggs, beat up with cream, thicken it over the fire, and add afterwards a dash of verjuice (or lemon)."*

"*Mussel Soup.*—Take the liquor that flows from the mussels when open in the fire, and strain it through a fine napkin; put it into some good broth; add the yolks of six eggs beat up with it, thicken it over the fire, and put it into your soup, when ready to serve, arranging the mussels round the dish."†

"*Mussel Fritters.*—Take them out of their shells, and steep them two hours in a quart of vinegar, some water, and a little butter rolled in flour, with salt, pepper, parsley, scallions, tarragon, garlic, a little carrot and parsnip, thyme, laurel, and basil; the whole make lukewarm, then take out your mussels, dry, and dip them in a batter, made with flour, white wine, and a spoonful of oil, and salt, and fry them."‡

The Neapolitans, as mentioned by Poli, eat mussels raw and fried, besides making patties and sauces of them.

"*Francatelli's Recipe for Scalloped Mussels.*—Scald and beard some dozens of mussels; strain the liquor into a stewpan, and add thereto 2 oz. of butter, mixed or

---

* 'French Family Cook.'   † 'French Family Cook.'
‡ Idem.

kneaded with 2 oz. of flour, a little cream, anchovy, nutmeg, and cayenne; stir the sauce over the fire to boil and reduce for ten minutes, then add a couple of yolks of eggs, a little lemon-juice, some chopped parsley, and add the mussels. Stir all together over the fire for a few minutes, and fill some scallop-shells with this preparation; cover them over with a thick coating of fried breadcrumbs, place them on a baking-sheet in the oven for five minutes, and serve them quite hot. They may also be served upon neatly-shaped pieces of dry toast."

"*Mussels dressed à la Provençale.*—Wash the mussels well several times, changing the water so as to cleanse them thoroughly; put them to dry in a saucepan over a hot fire, till the shells open. Take off one valve of the shell only. Put into a saucepan, half a glass of oil, parsley, chives, mushrooms, truffles, half a clove of garlic, all chopped very fine. Put it on the fire. Moisten it with a glass of white wine, a spoonful of broth, and half the quantity of liquor from the mussels. Boil this sauce, and when it is nearly reduced to half, add the mussels, with a spoonful of gravy; let the whole boil a few minutes; then add a spoonful of lemon-juice, pepper, and grated nutmeg,—then serve."*

"*Chilian Method of Cooking Shell-fish.*—A hole is dug in the ground, in which large smooth stones are laid, and upon them a fire is kindled. When they are sufficiently heated, the ashes are cleared away, and shellfish are heaped upon the stones, and covered first with leaves or straw, and then with earth. The fish thus baked are exceedingly good and tender, and this mode of cooking them is very superior to any other, as they retain, within

---

* Dictionnaire Général de la Cuisine Française ancienne et moderne.

the shell, all their own juiciness."* Meat dressed in the same manner is most delicious.

Cullis of Mussels.—Stew them and strain them; fry carrots, parsnips, parsley, basil, lemon, crumbs, a dozen of almonds; moisten them with broth; strain and keep the broth for use.

Mussels and Cockles must first be well washed in several waters, and then boiled in a closely covered saucepan, without water. When the shells open, take out the fish, strain the liquor, pick out the meat, carefully removing a tough membrane from the tongue of each mussel, and a substance resembling a small crab, which would be highly pernicious. To ascertain that nothing injurious remains, dip a silver spoon into the hot liquor. If it turns black, the next thing is *throw the whole away;* but, if otherwise, proceed to simmer the fish in the liquor, with a little salt and nutmeg, and a good piece of butter rolled in flour. Serve on toasted bread.

N.B.—This dish may be enriched by the addition of strong gravy, chopped mushrooms, anchovy, lemon-juice, and a larger proportion of butter, but it is generally preferred in its more simple form; indeed, many persons prefer having the fish served in the shells, to pick them out themselves, and eat with cold butter.†

Fam. OSTREADÆ.

OSTREA.—OYSTER.

Ostrea edulis, Linnæus. *Edible Oyster.*—Shell nearly round, though variously shaped, inequivalve; the

* King's 'Adventures of the Beagle,' vol. i. p. 291.
† 'The Housekeeper's Guide,' by Esther Copley, pp. 194, 195.

upper valve flat, or nearly so, with scales or laminæ of a yellowish-brown; the lower valve convex, and foliaceous, of a pale pinkish-white, with streaks of purplish-pink; transversely striated. Hinge toothless; ligament internal, of an olivaceous-brown; beaks small. The interior of the shell white and polished; sometimes *the purplish-pink colour of the margins showing through.*

The edible oyster of Great Britain is supposed to be superior to those of other European countries, and to attain to a greater degree of perfection on our coasts; and it was much valued by the Romans, who transplanted numbers from our shores, and placed them in artificial beds in the Lucrine Lake. Sergius Orata first invented these oyster-beds, "not for the gratification of gluttony, but of avarice, as he contrived to make a large income by this exercise of his ingenuity."* Apicius first discovered the art of preserving *oysters fresh* for a considerable time, and sent some from Italy to the Emperor Trajan, while he was on an expedition against the Parthians, which were found on their arrival to be as good as on the day they were gathered.† This mode may possibly have been the same as that which is practised in Italy at the present day, where, as Poli tells us, they are carried from Tarentum to Naples in bags, tightly packed with snow, which not only by its coolness preserves them, but also by preventing them from opening their bivalves, enables them to retain in the shells sufficient moisture to preserve their lives for a long period.‡

There were other places from whence oysters were procured, and Mucianus speaks with rapture of those

* Pliny, Nat. Hist. vol. ii. bk. ix. chap. 79.
† Daniel's 'Rural Sports,' vol. iv. p. 194.
‡ Poli, 'Testacea utriusque Siciliæ.'

found at Cyzicus, a town in Asia Minor,* on the shores of the Sea of Marmora, the ruins now called by the Turks, Bal Kiz. He describes them as larger than those of Lake Lucrinus; fresher than those of the British coasts; sweeter than those of Medulæ (the district in the vicinity of Bordeaux, now called Medoc); more tasty than those of Ephesus; more plump than those of Lucus; less slimy than those of Coryphas (a town of Mysia, opposite Lesbos); more delicate than those of Istria, and whiter than those of Circeii (a town of Latium). Pliny mentions that according to the historians of Alexander's expedition, oysters were found in the Indian Sea a foot in diameter,† and Sir James E. Tennent unexpectedly attested the correctness of this statement, as at Kottiar, near Trincomalee, enormous specimens of the edible oysters were brought to the rest-house. One shell measured more than eleven inches in length, by half as many broad.‡

The Greeks preferred the oysters of Abydos, and Archestratus, in his 'Gastronomy,' says:—

> "Ænus has mussels fine; Abydus too
> Is famous for its oysters; Parium produces
> Crabs, the bears of the sea, and Mityleue periwinkles;
> Ambracia in all kinds of fish abounds,
> And the boar-fish sends forth; and in its narrow strait
> Messene cherishes the largest cockles.
> In Ephesus you shall catch chemæ, which are not bad,
> And Chalcedon will give you oysters."§

Great Britain is still celebrated for its oysters, and large artificial beds are formed for the better rearing and breeding of these shellfish, besides the natural oyster-

---

* See Pliny, vol. vi. bk. xxxii. ch. 21 (6).
† Pliny, Nat. Hist. bk. xxxii. chap. 21 (6).
‡ See note, Nat. Hist. Ceylon, p. 371.
§ 'The Deipnosophists,' vol. i. bk. iii. p. 154.

beds, which are found on many parts of our coasts. The artificial beds require much labour to keep them in order, and free from shells and rubbish. The common mussel is an enemy to the oyster, as it causes mud to collect; and the starfish and whelk feed upon them, as do crabs and other shell-fishes. Dr. Paul Fischer states that the oyster-beds at Arcachon have suffered considerably from the havoc caused by *Murex erinaceus*, which has appeared in great numbers within the last ten years. It is known by the name of *cormaillot*, or *perceur*, and incessant war is waged against it; but the numbers do not decrease. Again, cold weather has a most pernicious effect upon the *spat*, for if the water is not warm enough the spat dies. Oysters will not even spawn if the weather is too cold.* Some of our principal beds are those of Whitstable, Rochester, Colchester, Milton, Faversham, Queensborough, and Burnham. Colchester has been celebrated for its oysters from a remote period, and they were deemed an appropriate present from the authorities of the town to ministers of state, and other eminent persons. We hear of their having been sent, in the reign of Queen Elizabeth, to Leicester and Walsingham.† At the annual Colchester Oyster Feast, held in the Townhall, October, 1862, Mr. Miller, M.P., mentioned that Mr. Goody, clerk to the Colne Fishery Company, with himself and a few other gentlemen, had appealed to the Treasury, because it was apprehended that Belgium, to which a large number of oysters are sent, was about to impose a duty which would inflict a serious injury upon the town. However, it was found from the interview that

---

* 'Times,' Mr. Buckland's Letter on Oysters, Aug. 3, 1864.
† Cromwell's Hist. of Colchester, vol. ii.

there was no immediate prospect of the anticipated danger, and a treaty was concluded with Belgium, in which a special reservation had been made in respect to oysters.* The oysters sent to Belgium are fattened in the Ostend beds, and then called "Ostend oysters;" though, in reality, they are *British*. They are very small and plump, and are highly thought of by the oyster-eaters in Paris; their flavour is certainly quite equal to the natives, at least I thought so, and the shells appeared thinner. Oysters, mussels, and periwinkles, with shrimps, are the fisheries which engage a good number of fishermen at Leigh, near Southend. The Leigh shore has been found particularly well adapted to grow and fatten oysters.†

Whitstable was a fishing-town of note in the reign of Henry VIII., and was called in ancient records "Northwood." Leland, in his 'Itinerary,' thus describes it:—"Whitstable is upward junto Kent, a ii miles or more beyond Faversham, on the same shore, a great fisher-towne of one paroche, belonging to Plaze College, in Essex, and yt standeth on the se-shore. Ther about they dragge for oysters."

The dredgers of Whitstable do not trust entirely to the natural resources of their oyster-beds, but purchase at Colchester or elsewhere, what is called the *brood*, which is the *spat* in its second stage. The following interesting account of the Whitstable beds appeared in 'Macmillan's Magazine,' No. 36, October, 1862:—"The brood is carefully laid down in the oyster-beds off Whitstable, and allowed to grow for three, perhaps four years. The oysters in different stages are marked

* 'Times,' October, 1862.
† 'Visits to the Seacoast; the Shipwrecked Mariner,' vol. xii. p. 30.

off by means of long poles, so that the shellfish farm is divided into separate fields, each being in a particular stage of growth. At the time when the oysters are lifted for the London or other markets, they are measured by being thrown against a wire grating, and all those under a certain size are thrown again into the water. To give an idea of the business done in the oyster trade, it may be stated that in 1860 the Whitstable men took as much as £50,000, for native oysters alone, which, after deducting the cost of the brood, would still leave a handsome profit." There are extensive fisheries opposite Milton, those of the Cheyney Rock. We are told that their farmer, Mr. Alston, has sent in a single season to London, more than 50,000 bushels of "natives" from this one fishery.* Mr. Buckland was informed upon the best authority, that out of the open sea no less than £180,000 worth of human food, in the form of oysters, is annually abstracted.†

The "Milton natives" bear the bell, or may be said to be the pearls among British oysters. King John granted these fisheries to the Abbot of Faversham, in whose hands they remained till the dissolution, and they have been dredged from the earliest times by a company of fishermen, ruled, like those of Faversham, by certain ancient customs and bye-laws.‡

Jersey oysters are brought over and bedded in Southampton Water, and the beds extend from thence to the coast of Osborne, in the Isle of Wight. They are described as being small, but of superior flavour, and are conveyed long distances to be laid down, naturalized, and afterwards sold as natives. They are also remark-

* Murray's Handbook, Kent and Sussex, p. 64. † See 'Times.'
‡ Murray's Handbook, Kent and Sussex, p. 64.

able for their saline flavour when first brought over, but it goes off after they have been bedded for some time at Southampton.*

It is said, that formerly there were fine oyster-beds between Portsmouth, Hayling, and the Isle of Wight, but they have been ruined by excessive and indiscriminate dredging.† I hear that there is still a considerable trade in oysters carried on at Hayling, and also at Brading. A bed of oysters was discovered off Eastbourne about three years since, the fish of a very superior and delicate flavour. The price was 1s. per hundred, but it has risen to 2s.,‡ and another large bed, which was valued at £5000, was found about three miles off the mouth of Dartmouth harbour, not very long ago.

The late Duke of Northumberland most successfully introduced oyster cultivation on the Northumbrian coast. They were imported and established there, and the year before last the fisheries were allowed to commence, when they were found to have succeeded admirably.

Messrs. Forbes and Hanley state that since the introduction of steamboats and railroads, considerable quantities of sea-oysters are brought from Falmouth and Helford, in Cornwall, also from Scotland and Ireland; the Irish oysters coming mostly from Carlingford, Malahide, Lissadell, Burran, Arklow, and Wexford. The Carlingford oysters are well known, and they are dredged by boats, each manned by from three to five men, who take about fifty dozen per day. Each boat deposits its oysters within a ring of small stones, till sold to dealers, the place being marked by a buoy. A yearly fee of 5s. is paid by each boat to the Marquis of Anglesey. The

* Illustrated London News.   † 'Field,' Note by Editor.
‡ 'Field,' February 27, 1864.

fishermen can earn from 4*d*. to 1*s*. 6*d*. *per diem*, and are mostly landholders. The Burran Bank oysters are highly esteemed in Dublin, and are called "Burton Bindons." They are brought from Kilkerran and Rossmuck Bays, in Galway, and are laid down to fatten on the Red Bank oyster-bed in Aughinish Bay. Formerly Mr. Burton Bindon was the possessor of these beds, but now Mr. Singleton has succeeded him, as we are informed by Mr. Buckland, who has recently visited these and other oyster-beds on the west coast of Ireland, the east coast of England, and also those on the west coast of France.

The Carrickfergus oysters are large in size, and, according to Mr. Patterson, so much in demand that their price in the Belfast market is generally from 12*s*. to 15*s*. per hundred of 120 oysters.

There are oyster-beds in the Shannon, said in 1836 to yield a revenue of £1400; and a small bed in Cork harbour, of no great extent, but the oysters are large, and prized for stewing, and sell at a good price. Formerly in Lough Swilly there were oyster-beds, and the oysters were sold at 2*d*. per hundred, but they have become very scarce, from the beds not having been properly protected.

In the 'Morning Post,' August 29, 1864, an account is given of the investigations by the commissioners appointed by Parliament, on the condition of the deep-sea fisheries of Ireland. At Wexford they elicited important facts. Fine oyster-beds are found at Wexford, but scarcely any oysters in them, which appears somewhat singular. It is not lawful to sell oysters in Ireland in the months of May, June, and July. The Wexford men dredge for them, of course, in the other months, but the reason of their beds being badly stocked is, that in the close months they are regularly dredged by Beaumaris boats, which replenish their own exhausted beds with them; and no later than last year, a French lugger visited Wexford seven times, carrying off on each occasion a large quantity of oysters for "laying down" on the French coast. Oysters are increasing in scarcity and dearness in Ireland; and, indeed, in our English markets also. This has led to much attention being given to oyster-culture, and various opinions expressed as to the most approved method of forming oyster-parks for the better rearing and preserving of these delicious bivalves. The "Fish and Oyster Culture" Company have established oyster-parks on the French principle at Prittlewell, and have laid down 1500 bushels of full-grown oysters, and it is calculated that the crop which will be secured in one year will be £1000 in value.

Between London and Glamorganshire there is a very large trade in pickled oysters.

In Scotland, the Cockenzie fishermen derive a good portion of their annual income from the oyster trade, and dredge for them at high and low tide. The crew of the boats keep up a wild and monotonous song (in which they believe there is much virtue) all the time they are dredging, and assert that it charms the oysters into the dredge.* The same authority further states, that as a class, the fishers of the Scottish coast are very superstitious. They do not like being numbered whilst standing or walking. It offends them very much to ask them whilst on their way to their boats, where they are going to-day; they consider it unlucky to see the impression of a very flat foot upon the sand, and they will not go

* 'The Fisher Folk of the Scottish East Coast,' Macmillan's Mag., No. 36, October, 1862.

to work, if in the morning on leaving their houses a pig should cross their path.

In the 'Sporting Gazette,' December 24th, 1864, an account is given of the discovery of a new oyster-bed in Glenluce Bay, Mull of Galloway. The oysters are large, and the fishermen think they are lying several feet deep. An experimental steam-fishing vessel has been built at Cockenzie; she is a dandy cutter-rigged craft, forty tons burden, assisted with auxiliary screw steam power, for the purpose of dredging oysters during the winter months, and deep-sea trawling during summer.

The celebrated "Pandore" oysters are principally obtained from Prestonpans. They are so called from being found in the neighbourhood of the salt-pans, and are large and well-flavoured.

Among the "Antient Cryes of London" we find the following:—

"We daily cryes about the streets may hear,
According to the season of the year;
Some *Welfleet oysters call*, others do cry
Fine Shelsea cockles, or white mussels buy."*

Oysters are also imported from the coast of Normandy and from Ostend; and Dr. Knapp tells us that not less than 800,000 tubs, each tub containing two English bushels, are procured annually from the Norman coast for the English market and the Channel Islands. Dieppe has large oyster-beds, and the celebrated *Rocher de Cancale* oysters are well known. There are fisheries at Granville, and the number of oysters taken in 1862 and 1863 was about 4,500,000; in 1861 and 1862 the Granville boats took 13,396,677 oysters, which, being sold at 18 francs the thousand, produced 241,140 francs.

* Kirby's 'Wonderful Museum,' vol. ii. p. 233.

The season of 1862–63, was not a profitable one; the oysters were sold at 28 francs the thousand, and the total profit of the season was 126,000 francs, the smallest amount yet known at Granville.*

An interesting paragraph appeared in the 'Times,' November 13th, 1862, on the cultivation of oysters on the western coast of France. It is as follows:—" M. Coste has just communicated a paper to the Academy of Sciences on the progress of his artificial oyster-beds. Several thousands of the inhabitants of the island of Ré have been for the last four years engaged in cleansing their muddy coast of the sediments which prevented oysters from congregating there, and as the work advances, the seed, wafted from Nieulle and other oyster-localities, settles in the new beds, and, added to that transplanted, peoples the coast; so that 72,000,000 of oysters from one to four years old, and nearly all marketable, is the lowest average registered per annum registered by the local administration, representing, at the rate of from 25 to 30 francs per thousand, which is the current price in the locality, a sum of about two millions of francs, the produce of an extremely limited surface. That the waves or currents carry the seed of oysters is a well-known fact, since the walls of sluices newly erected are often covered with them. In the island of Ré the existence of the oyster-beds, however, no longer depends upon this contingency, they being now in a state of permanent self-reproduction. Again, in some localities it is sufficient to prepare the emerging banks for collection, to see them soon covered with seed; but in other places nothing would be obtained without transplanting proper subjects. The concession

* 'Illustrated London News.'

of emerging banks is anxiously applied for by the inhabitants of the coast,—the more so, as improvements in the working of this branch of trade are of daily occurrence. Thus, Dr. Kemmerer, of Ré, covers a number of tiles with a coating of a kind of mastic, brittle enough to enable him to detach the small oysters from it. When this coating is well covered with seed, he gets it off all in one piece, which he carries to the place where the seed is to grow. The same tile he coats a second time, and so on."

In France, oysters having a green tint are considered great delicacies, and the art of greening oysters is carried to the greatest perfection on the coasts of Aunis, whence come the celebrated green oysters of Marennes. The oysters are, in reality, as white as the others, and only receive their green colour and peculiar flavour when transplanted to certain beds, covered with a small submarine kind of moss, formed of the slime deposited by the sea from the small gulf called the Rivière de Seudre. The 'Moniteur' published a letter from the Mayor of Marennes, in which he states that the trade in green oysters had increased so much during the last fifteen years, that the white oyster-beds in the neighbourhood had become insufficient to stock those peculiar beds where the creature acquires the green colour and that delicious taste which causes the Marennes oyster to be so eagerly sought after. In order to meet the demand, white oysters had to be imported from Spain, Bretagne, Ireland, and England. A considerable quantity of oysters are imported from Falmouth, and these contain copper, which imparts an acrid taste. They are generally, on their arrival, deposited in certain beds apart from the others, and there kept for six months; after which, experience has shown that they lose their copper, salt, and bad taste. A Marennes fisherman, whose trade was not very extensive, procured a few thousand oysters from Falmouth, and, out of thirst for gain, he sent them off to Rochefort before they had sojourned more than three weeks in the beds set apart for their purification.* These oysters caused alarming symptoms, and M. Cuzent, being called upon to test them, as they had been seized in the market at Rochefort, found copper in them, the quantity being about 23 centigrammes per dozen oysters. I have elsewhere given an account of the finding of copper in the Falmouth oysters; one of the tests used by M. Cuzent was so very simple, that any one might discover the presence of copper. It is as follows:—An ordinary needle is thrust into the green part of the oyster, and then the mollusk was immersed in pure vinegar. When copper was present, thirty seconds sufficed to cover the portion of the needle embedded in the oyster with a red coating of copper.†

The amount of shellfish consumed in Paris annually, including lobsters, crayfish, oysters, etc., is immense. Oysters are not packed in barrels, as with us, but at the restaurants and in the wine-shops are seen very shallow baskets, in shape resembling a small shield, with a thatching or wall of straw on either side, rising to the height of a foot or a foot and a half, tied with string at both ends and across the centre. These baskets contain a hundred or more oysters, according to their size.

In London, oysters are considered in season from the 4th of August to January, and the natives especially from October to March. The following were the prices at Billingsgate, August 4th, 1864:—Natives, 80$s.$ per

* 'Galignani's Messenger.'   † 'Field,' March 14th, 1863.

bushel; old royals, pearls, and Cheyney rocks, 30s.; other kinds averaging from 12s. 6d. to 14s., according to the quality. The shells of the native and rock oyster vary much, though they are only varieties of the same species, the shells of the latter being far more beautifully sculptured and coloured, though coarser and more rugged.

In the Bay of Cadiz, *Ostrea Virginica* is eaten when very small, but the poor people eat it full-size, viz. ten inches long. This species lives in the salt mud of the Guadalete, and is called *ostione;* other oysters are called *ostras*. The river is said to be salt three leagues from its mouth.

A Frenchman at Puerto St. Maria breeds oysters for the Madrid market, but they are slimy, and not to be compared with the English oysters, though they are said to be good when cooked; and Major Byng Hall* states that at Madrid, oysters—not fine ones—cost twopence halfpenny (that is, I suppose, one real) each; but this is not very remarkable, when in 1865 natives cost twopence, and Whitstable oysters three-halfpence each, in London, the very land of oysters, so scarce had these fish become.

The Tarentines declare that oysters are fattest during the full moon, and they are also fully persuaded that the moonbeams have a pernicious effect upon sea-fish; therefore they cover over fish taken by moonlight, lest they should decompose.

The American oyster, *Ostrea Virginica*, is much larger than the English, and differs from it both in taste and appearance. Four or six of them broiled are sufficient for a meal. Mr. Nichols, in his 'Forty Years in

* 'The Queen's Messenger,' p. 341.

America,' suggests acclimatizing it on our coasts;* and this has been done on the French coast, at Arcachon, and also at Saint-Vaast-la-Hogue. He tells us that oysters are never out of season at New York. They are brought from the shores of Virginia, and planted to grow and fatten, so that every quality and flavour can be produced by the varying situations of the banks, and the time of planting and the depth of water regulates the season of the oyster, and keeps the market in constant supply. It is not only in seaport towns in America that oysters are eaten in enormous quantities, but towns a thousand miles inland are well supplied; and oyster suppers are as common in Cincinnati or St. Louis as in New York or Baltimore.†

The amount of capital sunk in the oyster trade in the vicinity of New York exceeds £1,000,000.

Oysters are very beneficial to persons who suffer from weak digestions, but then they must be eaten raw, and without vinegar or pepper, and I have known an invalid able to eat oysters when quite unable to take any other food; and oysters are also recommended for consumptive patients, as they contain iodine.

The shells of the oyster and murex were used by the Romans as tooth powder, and oyster-shells are now used for manure.

Juan Francisco de San Antonio, in his 'Chronicos de

* In 1865, Sir Gardner Wilkinson found on the Tenby coast many fragments of the shells of *Ostrea Virginica*, and was led to suppose that this species of oyster (hitherto unknown or unnoticed in Britain) existed there; and he succeeded in obtaining some living and perfect specimens, the greatest number being met with in the neighbourhood of the small stream which runs into the sea on the south sands of Tenby. (See 'Zoologist,' 1865, p. 9558.)

† 'Forty Years in America,' vol. i. p. 268.

los Rel. Descalzos de S. Francisco,' etc., 1738, mentions the use of great oyster-shells for "holy water," and speaks of one known to be ninety years old, by the layers of its shell. But I fancy he must mean the shell of the *Tridachna gigas*, as we know it is used for that purpose; and in the church of St. Sulpice in Paris are two of these shells resting upon rock-work in marble, by Pigalle; they were given to Francis I. by the Republic of Venice. In the 'Intellectual Observer,' vol. i. p. 483, is an account of an "oyster-shell" island by M. Aucapitaine, on the east coast of Corsica, composed of layers of shells, bearing some resemblance to the shell-mounds of St. Michel-en-l'Herm, in La Vendée. This island is formed of still-living species, and is between three hundred and four hundred yards in circumference, the greatest elevation about thirty yards, and the mean elevation rather more than two yards above the level of the sea. The Romans are said by the fishermen to have deposited the shells of the oysters there, which they salted for exportation, but M. Aucapitaine does not believe in the artificial origin of this island.

According to M. de Quatrefages, the shell-mounds of St. Michel-en-l'Herm are composed of oyster, mussel, and scallop shells, of the same species as those living now in the neighbouring seas. Many of them have their valves still connected by the ligament which forms the hinge, and they have not even changed colour. The three banks of St. Michel-en-l'Herm are about seven hundred and thirty yards in length, three hundred in width, and rise about ten to fifteen yards above the level of the surrounding marshes.

Mr. Buckland mentions a large heap of oyster-shells in Galway Bay, at a place called Creggauns; another south-west of Tyrone, and one at Ardfry Point. The Creggaun heap consists principally of the shells of the oyster, mussel, and common cockle, though the whelk, *Pecten varius*, periwinkle, limpet, *Nassa reticulata*, *Helix nemoralis*, *Trochus*, and *Venerupis decussata* (*Tapes decussata ?*), are also found in it. There are also layers of wood-ashes and stones, apparently used as hearth-stones, showing the marks of having been subjected to fire, but no weapons. "The heap occupies an irregular space of two hundred feet long, and sixty feet wide, and ranges from six to eight feet deep." There are various traditions as to the age of the heaps; and it is said, that ninety years ago a series of high tides cast up the heap of shells from adjoining beds.*

At the present day the Baltic appears to be almost the only sea where the oyster will not grow,—a fact attributable to the very great influx of fresh water from the mouths of its many rivers, and the less powerful current from the ocean, so that, in the words of Sir Charles Lyell, "The *Ostrea edulis* cannot live at present in the brackish waters of the Baltic, except near its entrance." Yet, from the examination of the Danish Kjökkenmöddings, it appears, "that the oyster flourished in places from which it is now excluded, attaining its full size."

Oysters may be eaten in various ways, either cooked or raw.

> "The pepper-box, the cruet,—wait
> To give a relish to the taste;
> The mouth is watering for the bait
> Within the pearly cloisters cased.

---

* The 'Field,' February 4th, 1865.

> "Take off the beard,—as quick as thought,
> The pointed knife divides the flesh ;
> What plates are laden ! Loads are brought,
> Are eaten raw, and cold, and fresh." *

The oddest way of cooking an oyster, of which we have any mention, is that recorded by Evelyn,† who, in the year 1672, saw Richardson, "the famous fire-eater," perform wondrous feats, one of which was, "taking a live coal on his tongue, he put on it a raw oyster; the coal was blown on with bellows, till it flam'd and sparkl'd in his mouth, and so remain'd till the oyster gaped, and was quite boil'd." Who ate the oyster thus cooked, we are not informed.

"*Oyster Soup.*—Take 50 oysters; blanch them, but do not let them boil; strain them through a sieve, and save the liquor. Put ¼ lb. of butter into a stewpan; when it is melted, add 6 oz. of flour; stir it over the fire for a few minutes, add the liquor from the oysters, 2 quarts of veal stock, 1 quart of new milk; season with salt, peppercorns, a little cayenne pepper, a blade of mace, Harvey's sauce, and essence of anchovy, a tablespoonful each; strain it through a tammy, let it boil ten minutes; put the oysters into the tureen, with a gill of cream, and pour the boiling soup upon them."‡

Gower Recipe for Oyster Soup.—Boil 4 sheep's feet in 2 quarts of water, till reduced to 1 quart; it will then be a stiff jelly; put in it, while boiling, a small blade of mace; take off the fat, and thicken it with 1½ tablespoonful of ground rice; add from 20 to 50 oysters; boil it till thick enough, and add a teacupful of cream.

---

* Hone's 'Everyday Book,' vol. ii. p. 1071.
† 'Memoirs,' vol. i. p. 438.   ‡ Murray's 'Modern Cookery.'

Oyster Soup is also particularly good when made with a fish stock; as, for instance, with equal quantities of flounders, skate, and eels, or indeed with any fish that is abundant, and not much in request for other purposes.

"*Oyster Soup.*—Take 4 dozen oysters; lay the fish apart, and pass the liquor through a sieve, into a stewpan; set it on the fire; beat up the yolks of 6 eggs, and stir them in with half a pint of cream; add water or milk to the required quantity; season with pepper, a little grated lemon-peel, and the flesh of an anchovy beaten up, with a little butter and a small teaspoonful of good arrowroot. Five minutes before serving, put in the oysters."*

"*Potage à la Poissonière.*—Blanch 2 dozen oysters, 4 dozen of very fresh mussels, blanch and beard; put ¼ lb. of butter into a stewpan, with 6 oz. of flour, make a white *roux*; when cool, add the liquor of the oysters, mussels, and bones of a sole, with 2 quarts of broth, and 3 pints of milk; season with a spoonful of salt, one ditto of sugar, a sprig of thyme, parsley, 2 bay-leaves, 4 cloves, and 2 blades of mace; pass through a tammy into a clean stewpan; boil and skim well; cut about 10 pieces of salmon into thin slices, half an inch long, a quarter of an inch wide; cut the fillet of the sole the same size; put all into the boiling soup, with half a handful of picked parsley, and a gill of good cream; put the oysters and mussels in the tureen, and serve."†

"*White Oyster Sauce* (No. 43).—First scald and beard the oysters, and save their liquor. Next knead

---

* Maître Jacques.
† 'The Gastronomic Regenerator,' by Mons. A. Soyer.

2 oz. of butter, with 1 oz. of flour (or, better still, with arrowroot), in a stewpan; add the liquor, a gill of cream or milk, a little nutmeg, cayenne, anchovy, and lemon-juice; stir over the fire until the sauce boils, then add the oysters and serve hot."*

"*Brown Oyster Sauce* (No. 44).—Prepare the oysters as in the foregoing recipe, boil down their liquor, add half a pint of brown sauce (No. 12), or if there is none ready, use melted butter instead, adding a little browning; season with a little anchovy, cayenne, and lemon-juice; add the oysters; boil together for a few minutes, and serve hot."† Poli speaks of an oyster sauce made with honey,—or sugar,—vinegar, and various spices, but the mixture does not sound very inviting.

"*Oyster Sauce.*—Set a pint of cream upon the hob, beside a fire of clear glowing ashes, in an earthenware pipkin, glazed inside. Take 2 oz. of butter, and intimately mix with part of it a teaspoonful of best arrow-root, flavour with the flesh of an anchovy, pounded, a dash of cayenne-wine, a squeeze of lemon-juice, and a scrap of the peel, and stir in the whole, letting it boil until of the proper consistence; then put in the oysters, (if of a large size they should be cut into halves or quarters,) and keep stirring the sauce for about two minutes.—N.B. In mixing the butter with the cream, take care that the blending proceeds slowly, and keep stirring gently with a wooden spoon."‡

"*Oyster Atlets.*—Blanch throat-sweetbreads, and cut them into slices; then take rashers of bacon the size of the slices of sweetbreads, and as many large oysters blanched as there are pieces of sweetbread, and bacon.

* Francatelli's 'Cook's Guide.' † Ib.
‡ Maître Jacques.

Put the whole into a stewpan, with a piece of fresh butter, parsley, thyme, and eschalots, chopped very fine; pepper, salt, and lemon-juice, a small quantity of each. Put them over a slow fire, and simmer them five minutes. Then lay them on a dish, and when a little cool, put them upon a small wooden or silver skewer; a slice of sweetbread, a slice of bacon, and an oyster, and so on alternately till the skewers are full; then put breadcrumbs over them, which should be rubbed through a hair-sieve, and broil the atlets gently till done and of a light-brown colour. Serve them up with a little cullis under them, together with the liquor from the blanched oysters reduced and added to it."*

"*Curried Oyster Atlets.*—Take slices of sweetbreads, or slices of mutton or veal of the same size, put them into a stewpan with a piece of fresh butter, a tablespoonful of currie-powder, the juice of half a lemon, and a little salt. Set them over a slow fire, and when they are half-done, add to them blanched and bearded oysters, with their liquor free from sediment; simmer together five minutes, lay them on a dish, and when cold put them alternately on small wooden skewers. Then dip them in the liquor, strew fine breadcrumbs on each side, broil them over a clear fire till of a brown colour, and serve them up with some currie sauce under them.—N.B. The slices of sweetbread, oyster, veal, or mutton, to be of an equal number."†

"*Curried Oysters.*—Let a hundred of large sea-oysters be opened into a basin, without losing one drop of their liquor. Put a lump of fresh butter into a good-sized saucepan, and, when it boils, add a large onion, cut it into thin slices, and let it fry in the uncovered stew-

---

* Old Cookery Book. † Old Cookery Book.

pan until it is of a rich brown; now add a bit more butter, and two or three tablespoonfuls of currie-powder. When these ingredients are well mixed over the fire with a wooden spoon, add gradually either hot water, or broth from the stockpot, cover the stewpan, and let the whole boil up.

"Meanwhile, have ready the meat of a cocoa-nut, grated or rasped fine, put this into the stewpan with a few sour tamarinds (if they are to be obtained, if not, a sour apple, chopped). Let the whole simmer over the fire until the apple is dissolved, and the cocoa-nut very tender; then add a strong thickening made of flour, and water, and sufficient salt, as a currie will not bear being salted at table. Let this boil up for five minutes. Have ready also a vegetable marrow, or part of one, cut into bits, and sufficiently boiled to require little or no further cooking. Put this in with a tomata or two; either of these vegetables may be omitted. Now put into the stewpan the oysters, with their own liquor, and the milk of the cocoa-nut, if it be perfectly sweet; stir them well with the former ingredients; boil the currie, stew gently for a few minutes, then throw in the strained juice of half a lemon. Stir the currie from time to time with a wooden spoon, and, as soon as the oysters are done enough, serve it up, with a corresponding dish of rice on the opposite side of the table. This dish is considered at Madras the '*ne plus ultra* of Indian cookery.'"*

"*To Stew Oysters.*—Take the oysters clean from their liquor. Let the liquor stand till it is clear, then put a little of it to the oysters, and stew them; then put to

* Miss Acton's 'Modern Cookery Book,' taken from 'Magazine of Domestic Economy.'

them a little white wine, a little cream, and a little lemon-juice, a bit of butter, and shake them up together; then serve."*

"*Dutch Oysters.*—Roll rock oysters in yolk of egg; then dip them in grated breadcrumbs and white pepper, one by one, and fry them in butter. Serve with melted butter in a sauce tureen."†

"*To Fry Oysters.*—Take the largest oysters, open them, but do *not mangle them*, wash them in their own liquor, and take away all bits of shells; strew a very little flour over them. Dip them in the yolk of an egg, and fry them brown in butter."

"*Fried Oysters—Ostras Asada*, Spanish recipe.—Take the fish out of the shells, and simmer slowly for some minutes in their own liquor. Add salt, pepper, parsley chopped fine, a clove of garlic, some oil or butter, in which fry them gently; stir in a spoonful of flour, and moisten them with equal quantities of broth and wine. When done, add the juice of a lemon."

"*Fried Oysters; another way.*—Beat up two or three eggs in a cup, and rasp breadcrumbs on a plate, with sweet herbs powdered, and lemon peel. Dry the oysters as much as possible, souse them in the egg, and cover them with crumbs. Fry them in plenty of good butter, and serve with lemon-juice, cayenne, and brown bread and butter, cut thin."‡

"*A Ragoût of Oysters.*—Melt some butter, put in a little flour; keep it stirring till brown; wet it with gravy; put in a crust with the oysters and liquor; toss it; season with pepper, parsley, and fish broth."

---

* MS. Book.
† 'The English Cookery Book,' edited by J. H. Walsh, F.R.C.S.
‡ Maître Jacques.

"*A Ragoût of Oysters—Ostras Guisadas*, Spanish recipe.—Put the liquor of the oysters into a saucepan, with strong broth, and warm it; salt to your taste; then add the oysters, and a chopped anchovy or two; let them simmer, but not boil; serve with chicken, or white meat."

"*Grilled Oysters.*—Open and detach the largest oysters; place upon each a small piece of butter, well mixed with finely chopped parsley and spices; place them on the gridiron, and when they begin to boil serve them on a dish; or else, detach the oysters from their shells, and let them simmer in their liquor; take them out, and let them be placed again over the fire, with a piece of butter, parsley, some pepper, and a little lemon-juice. Put 4 oysters into each shell (after it has been well cleansed), and place the shell on the gridiron again for a few minutes, taking care not to let them boil up."*

"*To Roast Oysters* (206).—Place the oysters unopened between the bars of a fire, or in a charcoal stove. They require about six or eight minutes' time."†

"*Oysters—Ostras á la Pollada*, Spanish recipe.—Take oysters out of their shells, and blanch them in boiling water; then throw them into cold water, and take them out and let them drain. Put into a saucepan a piece of butter mixed with flour, parsley chopped fine, and mushrooms; warm this over the fire, and add sufficient broth to moisten it, and when it is thickened sufficiently, add the oysters, seasoned with pepper and salt, and let the whole boil. The moment before serving, add the juice of a lemon, or a little vinegar."

"*Oyster Sausages.*—Mince a pint of oysters, scalded so as to make them hard, and also a pound of lean sirloin

---

* La Cuisinière de la Campagne.   † The English Cookery Book.

of beef, and mix them; season with pepper, salt, and mace; mix up well with the yolks of 8 eggs; shape them like sausages, and fry in butter."*

"*To make Oyster Sausages.*—Take the flesh of the inside of a loin of mutton, and chop it as for force-meat, and season it with spice; then put to it 50 oysters, chopped very small, with a little French bread grated, and the yolks of 4 eggs, with a little chopped onion, a little beef-suet, and a little lemon-peel. *Roll* it into what form you please, and, if you do not use it, cover it up, and it will keep a long time."

"*To Mince Oysters.*—Take half a hundred of oysters, and put them into warm water; when they are ready to boil, shift them into cold water; then drain them, and take that part only that is tender. If you mix the flesh of carp with your oysters, it will increase your mince, and give it a better flavour. Put a bit of butter, shred parsley, scallions, and champignons, into a stewpan, and shake them over the fire, add a little flour, and moisten them afterwards with a gill of white wine, and as much soupe maigre; then put in your mince, and let it stew till the sauce be consumed; season it agreeably, and when you are ready to serve it, put in the yolks of three eggs, beat up with some cream."†

"*Oyster Force-meat.*—Open carefully a dozen of fine plump natives, take off the beards, strain their liquor, and rinse the oysters in it; grate 4 ounces of the crumb of a stale loaf into light crumbs, mince the oysters, but not too small, and mix them with the bread; add an ounce and a half of good butter, broken into minute bits, the grated rind of half a small lemon, a small saltspoonful of pounded mace, some cayenne, a little salt,

* Maître Jacques.   † The French Family Cook.

and a large teaspoonful of parsley. Mingle these ingredients well, and work them together with the unbeaten yolk of an egg, and a little of the oyster liquor, the remainder of which can be added to the sauce, which usually accompanies this force-meat."*

"*Scalloped Oysters.*—Scald and beard some dozens of oysters; strain the liquor into a stewpan, and add thereto 2 oz. of butter, mixed or kneaded with 2 oz. of flour, a little cream, anchovy, nutmeg, and cayenne; stir the sauce over the fire to boil, and reduce for ten minutes; then add a couple of yolks of eggs, and a little lemon-juice, and some chopped parsley; add the oysters, cut each into halves; stir all together over the fire for a few minutes, and fill some scallop-shells with this preparation; cover them over with a thick coating of fried breadcrumbs; place them on a baking-sheet in the oven for five minutes, and serve hot."† If you have no scallop-shells, the deep shell of the oyster, well scoured, will serve the purpose.

Many people, however, who prefer the real taste of the oyster, and do not like to conceal it beneath that of spice, prefer the old-fashioned way of scalloping oysters, which is as follows:—

"*Old way of Scalloping Oysters.*—Beard the oysters; scald the beards in the liquor from the fish, then strain them off; lay alternate layers of breadcrumbs, oysters, and small bits of butter in the shells, very slightly peppering them as you proceed. Pour the liquor in which you scalded the beards over them; put them into the oven till nicely browned, and if you find the colour not bright enough put them before the fire for a few minutes, or salamander them. A little cream, added after the

* Miss Acton's 'Modern Cookery.'   † Francatelli's 'Cook's Guide.'

shells are filled, but before they are put in the oven, is a great improvement."

By lining the dish, and covering the oysters with puff paste, this is converted into an *Oyster Pie*, which makes an excellent dish.

"*Scalloped Oysters—Ostras en Concha*, Spanish recipe. —Select the largest shells, and scrub them very clean; put 4 or 6 oysters into each, with their liquor, and cover them with breadcrumbs, seasoned with pepper and salt; then place the shells on the gridiron till the fish is cooked."

"*Oyster Fritters* (2997).—Make a batter of flour, milk, and eggs; season with a very little nutmeg. Beard the oysters, and put as many as you think proper in each fritter."*

"*Oyster Loaves.*—Open the oysters, and save the liquor; wash them in it; then strain it through a sieve, and put a little of it into a tosser, with a bit of butter and flour, white pepper, a scrape of nutmeg, and a little cream, stew them, cut in dice; put them into rolls sold for the purpose."†

"*Oyster Pie.*—As you open the oysters separate them from the liquor, which strain; parboil them, after taking off the beards; parboil sweetbreads, and cutting them in slices, lay them and the oysters in layers; season very lightly with salt, pepper, and mace; then put half a teacupful of liquor, and the same of veal gravy. Bake in a slow oven; and before you serve put in a teacupful of cream, a little more oyster liquor, all warmed, but not boiled."‡

Oyster Pie.—Take a large dish, butter it, spread a

* 'Enquire Within upon Everything.' † The English Cookery Book.
‡ Murray's 'Modern Domestic Cookery.'

rich paste over the sides, and round the edge; but not at the bottom; the oysters should be fresh, and as large and fine as possible; drain off part of the liquor from the oysters; put them into a pan, season them with pepper, salt, and spice; stir them well with the seasoning; have ready the yolks of eggs, chopped fine, and grated bread; pour the oysters (with as much of their liquor as you please) into the dish that has the paste in it; strew over them the chopped egg and grated bread; roll out the lid of the pie and put it on, crimping the edge handsomely. Bake the pie in a quick oven.

"*Pickled Oysters.*—Put 2 dozen of large oysters into a stewpan over a fire, with their liquor only, and boil them five minutes; then strain the liquor into another stewpan, and add to it a bay-leaf, a little cayenne pepper, salt, a gill and a half of vinegar, half a gill of ketchup, a blade of mace, a few allspice, and a bit of lemon-peel; boil it till three parts reduced, then beard and wash the oysters, put them to the pickle, and boil them together two minutes. When they are to be served up, place the oysters in rows, and strain the liquor over them; garnish the dish with slices of lemon or barberries."*

Glamorganshire way of Pickling Oysters. — Beard them nicely; then slowly stew them in the liquor from their shells, with a bay-leaf or two, and some whole black pepper; a very small quantity of vinegar is then added, and they are placed in stone jars, corked, and covered with pitch. They are then ready for the London markets.

This oyster pickling may be seen going on in almost every cottage. The oysters when raw sell at 1*s.* a

* From an old Cookery Book.

hundred, and when pickled at about 1*s.* 9*d.*, or even at 2*s.*

Soyer's Recipe for Pickling Oysters for the London Markets.—" Put the oysters, with their liquor, in an earthen pan on the fire to simmer; take off the scum as it rises; add some whole pepper, sliced ginger (green if possible), a few cloves, some chopped chilies, and a little vinegar; simmer not longer than five minutes, and take them out; remove the beards, and put the oysters in a barrel, and when the liquor is cold, strain and add it."

Pickled Oysters—Ostras en Escabechados, Spanish recipe.—" Make a pickle of the liquor of the oysters, chopped onions, parsley, garlic (this, of course, may be omitted if not liked), bay-leaves, marjoram, salt, pepper, butter into which flour has been rubbed, and a few drops of vinegar; when well thickened by boiling, add the oysters, and stir gently."

"*Oyster Powder.*—Open the oysters carefully, so as not to cut them, except in dividing the gristle which attaches the shells; put them into a mortar, and when you have got as many as you can conveniently pound at once, add about 2 drachms of salt to about a dozen oysters; pound them, and rub them through the back of a hair-sieve, and put them into a mortar again, with as much flour (but previously thoroughly dried) as will roll them into a paste; roll this paste several times, lastly flour it, and roll it out the thickness of a half-crown, and cut it into pieces about one inch square; lay them in a Dutch oven, where they will dry so gently as not to get burned; turn them every half-hour, and when they begin to dry crumble them; they will take about four hours to dry; pound them, sift them, and put them into dry bottles; cork and seal them. Three dozen of natives require 7½ oz. of flour to make them into a paste weighing 11 oz. and when dried 6½ oz. To make half a pint of sauce, put 1 oz. of butter into a stewpan, with 3 drachms of oyster powder, and 6 tablespoonfuls of milk; set it on a slow fire, stir it till it boils, and season it with salt; as a sauce, it is excellent for fish, fowls, or rump-steaks. Sprinkled on bread-and-butter it makes a good sandwich."*

"*Oyster Ketchup.*—Pound the fish, and add to each pint of them, 1 pint of sherry wine, 1 oz. of salt, powdered mace 2 drachms, pepper 1 drachm. Boil up, skim, strain; add to each pint, brandy 2 teaspoonfuls, then bottle. To flavour sauces when oysters are out of season."†

"*Oysters au Gratin.*—Set a little cream in a pipkin, with a piece of butter (the quantities to be judged according to the size of the dish), and mingle them gradually; add to this a little anchovy sauce, cayenne, wine, and grated lemon-peel. Pour half of this in a dish, lay in the oysters, and grate over them a little Parmesan cheese and breadcrumbs (not too thick a layer), seasoned in the usual way; then pour over the rest of the cream and butter, and grate another thin layer of Parmesan and breadcrumbs. Set it in a quick oven, or in a Dutch oven."‡

* 'Enquire Within upon Everything.'
† 'Dictionary of Practical Receipts,' by G. W. Francis, F.L.S.
‡ Maître Jacques.

Fam. PECTINIDÆ.
PECTEN.—SCALLOP.

PECTEN OPERCULARIS, Linnæus. *Lid Scallop.* — Shell spherical; valves convex, of nearly equal dimensions, rather strong; ribs 18 or 20 in number, finely striated, both longitudinally and transversely; auricles nearly the same size; ligament internal; hinge without teeth.

This is the common scallop of the people, and much smaller than the "great scallop," also subject to greater variety of colour. Specimens are found quite white, with a dark red line on the summit of each of the radiated ribs, (var. *lineatus,*) also brown, yellow, speckled white and brown, purplish-pink, and orange. The specimen figured was dredged up off the Parson and Clerk rocks, at Dawlish, and at times there may be gathered baskets full on the beach between that town and the mouth of the Exe. The shells are much used in ornamental work; and pretty baskets, pincushions, needle-books, etc., are made from the beautiful variegated valves.

The scallop may be called the butterfly of the ocean, from its power of swimming or flying rapidly through the water. This was observed by Pliny, who says that the scallop is able to dart above the surface of the water, just like an arrow.* By some this power is supposed to be caused by the rapid opening and shutting of the valves; but Mr. Gosse states that after carefully watching the habits of a Pecten, which he kept for some days in a glass phial of sea-water, he discovered that the flitting motion was performed by forcing jets of

* Pliny, Nat. Hist. vol. ii. bk. ix. ch. 45 (29).

water through the compressed edges of the mantle. He says, "When the Pecten is about to leap, it draws in as much water as it can contain within the mantle, while the lips are held firmly in contact. At this instant the united edges of the lips are slightly drawn inward, and this action gives sure warning of the coming leap. The moment after this is observed, the animal, doubtless by muscular contraction, exerts a strong force upon the contained water, while it relaxes the forced contact of the lips at any point of the circumference, according to its pleasure. The result is, the forcible ejection of a jet of water *from that point,* which, by the resilience of its impact upon the surrounding fluid, throws the animal in the *opposite direction,* with a force proportioned to that of the *jet d'eau."* Again, Mr. Gosse adds, "That the Pecten widely opens and forcibly closes its valves if *left uncovered by the water,* is, doubtless, correct. I have seen my specimen perform such an action, and perhaps it might by such means jerk itself from place to place, with considerable agility. But I do not think so rude a mode of progression could enable it to select the direction of its leaps, which under water appears to me to be determined with so much precision."*

Scallops are found pretty generally distributed in all seas, and are much sought after for food. At Weymouth, the average produce of the trawlers is five bushels of scallops per week. They have been sold at 2*d.* per hundred, 700 going to the bushel;† but they appear to have become scarcer lately, if one may judge by the price at which they are now sold, viz. 4*d.* a dozen, and 2*d.* per dozen for the shells without the fish, for making into shell

* 'Devonshire Coast,' by P. H. Gosse, pp. 50 and 52.
† 'A Year at the Shore,' by P. H. Gosse, p. 25.

ornaments, and the fishermen suppose that they are taken in the greatest numbers after a fall of snow. In Cornwall they are caled *frills*, or *queens*; on the Dorset coast *squinns*, and in the north of France, *vanneau* or *olivette*;* and in the south of Ireland, the peasantry call them *closheens*. *Pecten varius* is sent in quantities from the department of Charente Inférieure to the market at Bordeaux, and is called *la petite palourde*,† and in the north of France, *petite vanne*; and according to Poli it is the *pellerinella* of the Neapolitans, and the *canestrelli di mare* of the Venetians.

"*To Fry Scallops.*—Wash the shells well in clean water, then put them into a saucepan over a slow fire until they are open; then take out the fish and place them on a dish, covering them well with breadcrumbs or flour, and add a little pepper. Then put some oil, lard, or butter into a frying-pan, and when it begins to boil, put in the scallops and fry them till they are well browned. Shake the frying-pan occasionally, to prevent their mixing together."

Soyer, in his 'Ménagère,' gives the following recipe: —" Escallop is exceedingly fine; it should be kept in salt and water some time, to free it from sand. When opened, remove all the beard, and use only the white, red, and black parts. It may be cooked like oysters, and is excellent with matelote sauce."

In Francatelli's 'Cook's Guide' is a recipe for oyster soup, but he adds that a good soup may be made in the same manner, substituting scallops instead of oysters, and I shall therefore give it:—

"*Oyster Soup (Scallop Soup*, No. 183).—Scald, drain,

---

* Jeffreys, 'British Conchology,' p. 60, vol. ii.

† 'Faune Conchyliologique Marine,' etc., par le Dr. Paul Fischer.

wash, and beard 4 dozen of oysters (or scallops), reserving their liquor in a pan. Put 4 oz. of butter into a stewpan, to barely dissolve over the fire; mix in 4 oz. of flour; moisten with a pint and a half of good white stock or milk; season with nutmeg, and a pinch of cayenne, and a teaspoonful of anchovy; add half a pint of cream; stir over the fire for a quarter of an hour's gentle boiling, and then, having cut the oysters (or scallops) each into halves, pour the hot soup over them in the tureen."

"*To Cook Scallops, or 'Leitrigens,' Donegal fashion.*—Place them on a gridiron in the shells, with a piece of lighted turf-coal placed on the upper shell; when cooked, eat them with butter and pepper."

Gwillim, in his 'Heraldry,' says that (according to Dioscorides) the escallop is "ingendred of the dew and the air, and hath no blood at all in itself; notwithstanding, in man's body, (of any other food) it turneth soonest into blood," and adds, "the eating of this fish raw is said to cure surfeit."

PECTEN MAXIMUS, Linnæus. *Great Scallop.*—Shell suborbicular; valves very dissimilar, the upper one concave at the umbones; the under valve very convex; strong ribs, 15 or 16 in number; rather broad, and distinctly striated; auricles large, nearly equal; hinge without teeth; ligament internal, placed in a triangular recess.

The great edible scallop, though generally distributed in our seas, is only locally abundant. At Eastbourne and Brighton numbers are brought in by the fishing boats, and in the spring, during the prevalence of the easterly gales, live specimens may be found on the beach at Dawlish. The London markets are supplied from various parts of our coast, but I am told that tons of scallops and periwinkles are sent yearly from Brading harbour, in the Isle of Wight; but the greatest supply is from Holland. They are sold at 2s. per dozen, and are chiefly sought after for the shell. There are large scallop beds off the Isle of Man. At Vigo, *Pecten maximus* is the constant food of all classes from Christmas to Easter, after which it is only eaten by the very poor people, and there it is known by the name of *Beíra*. In Andalusia it is called *Rufina*.

The French call the scallops *peignes, coquilles de St. Jacques*; also, *grosille, grand'-pèlerine, gofiche*, or *palourde*,* and the name for them in German is *Jacobsmuschel, Pilgrims-muschel*, and *Kamm-muschel*. At Tarento, the fishermen call this shell *Concha di San Dialogo*, and in other places, *Cappa di San Giacomo*, and consider it a great delicacy; and formerly it grew so large there, that Horace says:—" Pectinibus patulis jactat se molle Tarentum."† According to Poli, the Neapolitans call it *Cozza di San Giacomo*, and the Venetians *Cappa Santa*. In Youghal these mollusks are known by the name of *kirkeens*, or *kirkeen thraws*; another Irish name for them is *Sligane-mury*. In Scotland, scallops are often called *clams*, and are used as bait for the white-fish lines, but other shells are called *clams*; amongst them is *Pholas dactylus*, which is generally used by us as bait, though eaten in France;‡ and in the Shetland Isles the large *Cyprina Islandica* is the *clam*. A species of *Mya*, eaten by the natives of the Zaire or Congo River, is stated by Mr. Fitzmaurice to resemble what is usually called the *clam* in England; and at

* Jeffreys, 'British Conchology,' vol. ii. p. 74.
† Aufrère's Travels.   ‡ 'Book for the Seaside,' p. 84.

Dawlish, the *Solen* is called the *sand clam*. *Lutraria maxima* is called the *great clam*, or otter shell, and Mr. J. K. Lord states that in British Columbia and Vancouver's Island it is one of the staple articles of winter food, on which the Indian tribes, who inhabit the North-West Coast of America, in a great measure depend. The squaws fish for them, as it is derogatory to the dignity of a man to dig *clams*. They use a bent stick for the purpose, about four feet long, and they cook them by placing the shells on red-hot pebbles from the camp fire, till the shells open. To preserve them for winter use, a long wooden needle with an eye at the end, is threaded with cord made from native hemp, and on this the clams are strung like dried apples, and thoroughly smoked in the interior of the lodge.* The wampum, or Indian money, is made of the clam (*Venus mercenaria*), and the shells are strung together and form a belt. The wampum is the token of peace and friendship amongst the American Indians.

Professor Simmonds, in his 'Curiosities of Food,' gives an account of the collection of clams on the North American coasts for the Boston markets. There are two kinds, which are eaten in great numbers in spring, when they are in the best condition. They are also salted and preserved in barrels, and the fishermen use them as bait for codfish. It has been suggested by Mr. Nichols to acclimatize the American clam on our British and Irish coasts, as it would prove a valuable addition to our edible mollusks,† and the experiment has already been tried on the French coast by M. de Broca, M. Coste, and Count de Férussac. Breeding-beds

---
* 'The Naturalist in British Columbia,' by John Keast Lord, vol. i.
† 'Forty Years in America,' Nichols.

were prepared on the coast at Arcachon and Saint-Vaast-la-Hogue, and in 1861, the steward of the 'Arago' steamer brought over about two hundred clams, and also some American oysters, which were deposited in these beds under the superintendence of M. Coste.* In 1863, another supply of live clams was brought over; but Dr. Paul Fischer stated in 1865, that though the mollusks were perfectly healthy, they did not seem to have spawned, as no young specimens could be found. This *Venus mercenaria* is a very thick shell; covered with a dark-brown epidermis, much resembling our *Cyprina Islandica*, but it is more triangular in form.

The deep valves of *Pecten maximus* are used by fishermen as lamps for their huts, and, according to Fuller, they were also made use of by the pilgrims in Palestine, as cups and dishes, but I believe that the real pilgrim scallop is *Pecten Jacobæus*, which is found in the Mediterranean, and is smaller, of a very bright colour, dark-orange and reddish-brown, more convex, the ribs more defined and angular. The scallop was also the badge of the pilgrim, and the poet Bowles says:—

> "He clad him in his pilgrim weeds,
>   With trusty staff in hand
> And scallop shell, and took his way,
>   A wanderer through the land."

Again, in 'Marmion,' we read:—

> "The summoned Palmer came in place,
> His sable cowl o'erhung his face;
> In his black mantle was he clad,
> With Peter's keys in cloth of red
>   On his broad shoulders wrought;
> The 'scallop shell' his cap did deck;

---

* 'Utilization of Minute Life,' by Dr. T. L. Phipson, pp. 176, 177.

> The crucifix around his neck
> Was from Lorotto brought ;
> His sandals were with travel tore,
> Staff, budget, bottle, scrip he wore :
> The faded palm-branch in his hand,
> Showed pilgrim from the Holy Land."

At the present day, many distinguished families bear scallop shells on their shields, showing that their ancestors had made pilgrimages to the Holy Land, or other distant shrines; and Fuller says:—

> "For the scallop shows a coat of arms,
> That, of the bearer's line,
> Some one in former days hath been
> To Santiago's shrine."

The scallop shell may be seen in the arms of the Duke of Bedford, the Earl of Jersey* (whose ancestor, Sir Richard de Villars, " assumed the coat of arms, *argent, on a cross gules five escallops or*, in the reign of Edward I., as a badge for his services in the Crusades), the Marquis Townshend, Lord Dacres, and many others. An escallop argent, between two palm-branches vert, is the crest of Bullingham, of Lincolnshire; and that of Bower, of Cloughton and Bridlington, Yorkshire, is an escallop argent.

The arms of Buckenham Priory, Norfolk, founded about 1146, by William de Albini, Earl of Arundel, and Queen Adeliza, his wife, widow of King Henry 1., were *argent, three escallops sable ;* and the seal of the Priory bears the figure of St. James as a pilgrim, with the scallop shell in his hat, a pilgrim's staff in one hand, and a scrip in the other.† Another old abbey seal, of which I have seen the impression, has the figure of St. James

---

* 'The Noble and Gentle Men of England,' by E. P. Shirley, Esq.
† Moule's 'Heraldry of Fish,' p. 223.

(or Saint Jacques de la Hovre) in his pilgrim's dress, his staff in one hand and a scrip in the other, with a scallop shell on either side of the figure. The inscription, unfortunately, I could not read, as it was indistinct. On many monumental slabs and tombs the scallop shell appears; and in Melbourne Church, Derbyshire, in a canopied recess in the chancel, is a recumbent figure of a knight, or crusader, with mail and surcoat, with a shield on his arm bearing three scallop shells, with chevron between. The monument is much mutilated, and it is not known to whom it belongs. Again, in St. Clêment's Church, Sandwich, is a slab with the date 1583, to the memory of "George Raw, gent., sometyme mayor and customer of Sandwic, and marchant adventurer in London ;" with a shield bearing the arms, ermine on a chief (gules) two escallop shells (or); crest, a dexter arm embowered in armour (sable), garnished (or), holding a scallop shell. However, the escallop in heraldry is borne not only as a badge of pilgrimages, but by those who have made long voyages, have gained great victories, or have had important naval commands.*

It is curious to remark, that leaden coffins ornamented with scallop shells, rings, and beaded pattern, belonging to a much earlier period, have been dug up from time to time on the sites of Roman cemeteries. Mr. C. Roach Smith, in an interesting paper on 'Leaden Coffins,' in 'Journal of the Archæological Association,' vol. ii., mentions several. Two were found at Colchester, and near one of them was an urn, in which were two coins, one of Antoninus Pius, and the other of Alexander Severus; again, in Weever's 'Funeral Monuments,'

* See 'Crests of Great Britain and Ireland,' vol. i. p. 525, by Fairbairn.

mention is made of a similar coffin (discovered in the parish of Stepney, Middlesex, in the district known to occupy the site of one of the cemeteries of Roman London), the upper part ornamented with scallop shells; having at the head and foot two jars; on the sides a number of bottles of glistening red earth, some of which were painted, and also some glass phials. The chest, or coffin, contained the body of a woman. Leaden coffins have been found at York, and in a Roman tomb at Southfleet, Kent, and other places, as well as in France; and Mr. C. Roach Smith says, "that they may, most of them, possibly be assigned to the Roman-British period."

The scallop shell appears legitimately to have belonged to pilgrims to the shrine of St. James of Compostella, as may be gleaned from the following legend, given by old Spanish writers:—

"When the body of the saint was being miraculously conveyed in a ship without sails, or oars, from Joppa to Galicia, it passed the village of Bonzas, on the coast of Portugal, on the day that a marriage had been celebrated there. The bridegroom with his friends were amusing themselves on horseback on the sands, when his horse became unmanageable, and plunged into the sea; whereupon the miraculous ship stopped in its voyage, and presently the bridegroom emerged, horse and man, close beside it. A conversation ensued between the knight and the saint's disciples on board, in which they apprised him, that it was the saint who saved him from a watery grave, and explained the Christian religion to him. He believed, and was baptized there and then, and immediately the ship resumed its voyage, and the knight came galloping back over the sea to rejoin his astonished friends. He told them all that had happened, and they, too, were converted, and the knight baptized his bride with his own hand. Now, when the knight emerged from the sea, both his dress and the trappings of his horse were covered with scallop shells; and, therefore, the Galicians took the scallop shell as the sign of St. James."*

Florez† says that a Galician peasant discovered, in the ninth century, the spot in which was deposited a marble sepulchre, containing the ashes of St. James, owing to the appearing of certain preternatural lights in a forest. The shells of Galicia, or scallops, belonged exclusively to the Compostella pilgrim, and the Popes Alexander III., Gregory IX., and Clement V., in their Bulls, granted a faculty to the archbishops of Compostella, to excommunicate all who sold these shells to pilgrims anywhere except in the city of Compostella.‡

When the marriage of Edward I., King of England, took place with Leonora, sister of Alonzo of Castile, a protection to English pilgrims was stipulated for, but they came in such numbers that they alarmed the French, who threw difficulties in their way. In the fifteenth century, Rymer mentions that 916 licences were granted to make the pilgrimage to Santiago in 1428; in 1434 as many as 2460 were granted.§ The name of Jacobitæ, or Jacobipetæ, was given to Compostella

* "Pilgrims of the Middle Ages," Rev. E. L. Cutts, M.A., 'Art Journal,' 1861, p. 309.

† 'Historia Compostellana,' lib. i. cap. ii. apud 'España Sagrada,' tome xx.

‡ "On Pilgrims' Signs and Tokens," C. Roach Smith. See note, Archæological Journ., vol. i. p. 202.

§ See note, "Pilgrims of the Middle Ages," vol. vii. p. 308, Rev. E. L. Cutts.

pilgrims, and there was an hotel at Paris on purpose for receiving them if they were bound to St. James's shrine; but the revenues failing, it was purchased by the Dominicans.* "Besides its badge, each pilgrimage had also its gathering cry, which the pilgrims shouted out, as at grey of morn they slowly crept through the town or hamlet where they had passed the night,"† and Pope Calixtus‡ says that the Santiago pilgrims were accustomed, before dawn, at the top of each town, to cry with a loud voice, "Deus adjuva! Sancte Jacobe!"§

It is stated that pilgrims used to present their scrips and bourdons to their parish churches, and Coryatt saw cockle, mussel-shells, beads, and other religious relics, hung up over the door of a little chapel in a nunnery. These were deposits and offerings made by pilgrims to Compostella, when they returned and gave thanks.‖

The Rev. E. L. Cutts states that shells have not unfrequently been found in stone coffins, and are supposed to be relics of the pilgrimage once taken by the deceased to Compostella; and that when the grave of Bishop Mayhew, who died in 1516, was opened some years ago, in Hereford Cathedral, a common rough hazel-wand, between four and five feet long and as thick as a man's finger, was found lying by his side, and with it a few mussel and oyster shells.

St. James of Compostella is said to have performed many miracles, and to have appeared no less than fifteen several times to the Spanish kings and princes,

* Fosbroke's Brit. Monachism, p. 469.
† "Pilgrims of the Middle Ages," p. 321, 'Art Journal.'
‡ See note, "Pilgrims of the Middle Ages." Sermones, Bib. Pat., ed. Bignio, xv. 330.
§ Dr. Rock's 'Church of the Fathers,' vol. iii. p. 442.
‖ Brit. Monachism.

when some great advantage always ensued; for instance, one day he put himself at the head of the troops of a King of Spain, Ramira, King of Leon, and leading them against the Moors, mounted on a white horse, the housings charged with escallops,* defeated those infidels. St. James supported his people, by taking part in their battles, down to a very late period, as Caro de Torres mentions two engagements in which he cheered on the squadrons of Cortes and Pizarro "with his sword flashing lightning in the eyes of the Indians."† The great Spanish military order of Santiago de la Espada is supposed to have been instituted in memory of the celebrated battle of Clavijo; the peculiar badge of which order is a red cross, like a sword, charged with a white scallop shell, and the motto, "Rubet ensis sanguine Arabum."‡ To this day you are told in Spain, that the scallops found at Clavijo were dropped there by St. James, or Santiago, when he assisted the Spaniards to kill 60,000 Moors, in the year 997, and they are considered visible proofs for those who doubt the miracles of this saint.

Other orders of knighthood used the scallop shell as an ornament, viz., that of St. James of Holland, the badge and collar being formed of escallops; and Louis IX. of France, or St. Louis, as he was generally called, instituted an order of knighthood, called the "Ship and Escallop Shell," to induce the French nobility to accompany him in his pilgrimage to the Holy Land.§ He quitted Paris the 12th of June, 1248, to embark at

* 'Heraldry of Fish,' p. 222.
† 'Ordenes Militares,' fol. 5. Note, Prescott's 'Ferdinand and Isabella,' vol. i. p. 274.
‡ Heraldry of Fish. § Ibid.

Aigues-Mortes, in Languedoc; a town which he had founded, that he might have a seaport on the Mediterranean. He also embarked at that place on his unsuccessful crusade in 1270, having assembled a fleet of 800 galleys, and an army of 40,000 men.

The following description of the apostle St. James, patron of Spain, given by Bernard Picart,* may not be uninteresting to some of my readers. He says, " St. James, Patron of all Spain, has rested for these 900 years past in the Metropolitan Church of Compostella. The image of this blessed apostle is upon the high altar; it is a small wooden bust, with forty or fifty white tapers constantly burning before it. Pilgrims kiss it three times, and put their hats upon the head of it, with abundance of respect and devotion. There are thirty silver lamps always burning in the church, and six large silver candlesticks five feet high, which were given by Philip III. There are five platforms, of large freestones, for walking all round the church, and above it is another of the same kind, where the pilgrims ascend and fix some remnant of their clothes to a stone cross, which is erected thereon. They likewise perform another ceremony as singular as this. They pass under this cross three times, through such a small hole that they are obliged to slide through with their breasts against the pavement, so that such as are never so little too fat must suffer severely, and yet through they must go, if they will obtain the indulgence thereto affixed. This is the strait gate of the Gospel through which the pilgrims enter into the high-road to salvation. Some who had forgotten to pass under the stone cross, have gone back five hundred leagues to perform this ceremony."

* 'Religious Ceremonies,' p. 432.

Mr. Street, in his 'Gothic Architecture in Spain,' states that, even in that country, the old belief of the power of the bones of St. James of Compostella to work miracles appears now practically to have died out, and that there are no longer great pilgrimages to his shrine. However, at Santiago de Compostella he saw *one professional* pilgrim with his rags covered with scallop shells, whom he had previously seen begging at Zaragoza; and in one of the Plazas at Santiago, an old woman was selling scallop shells.

The custom of bearing scallop shells as a badge of pilgrimage, is more widely spread than is usually supposed, for Sir Rutherford Alcock mentions their use on the sleeves of many of the Japanese pilgrims to the Cone of Fusiyama, in the island of Japan.

Shells were used by the Romans to ornament their dwellings, and the 'Fountain of Shells,' described in Sir W. Gell's 'Pompeiana,' was decorated with the Tyrian Murex and the scallop.*

The scallop is figured on the coins of Saguntum, which are of Phœnician time, the dolphin being on one side, with the letters S. A. G. W. under, and the scallop on the reverse; and Florez, in his 'Medallas de España,' Parte 2, 1728, says of these coins: " These (the dolphin and the scallop shell) allude to Neptune and Venus, for as the dolphin is sacred to Neptune, so is the shell to Venus,† as the daughter of the sea, and also for the pearls it engenders, applied to the adornment of women. This shell is most appropriate for the impress of a maritime city, from the utility enclosed within it, and its application to diverse uses, either from its seed for jewels,

* Jeffreys, Brit. Conchology, vol. i. p. 67; Introduction.
† "Faveas concha Cypria vecta tua," Tibullus, lib. iii. El. 3, etc. etc.

or as a delicacy for the table, for the precious tints with which it is coloured, for its use as a medicine, and for ostentation in virtue of its ornamental pearls."

Real scallop shells are also used in the baptismal service for pouring water over a child, though the shell is generally of silver gilt, and in private baptism a wooden shell is frequently adopted. "Baptismal shells" are mentioned in a list of the ornaments of the Church in the fifteenth century, and they are still used in some churches.

The following are a few recipes for cooking the scallop :—

"*To dress Scallops.*—Wash them six or seven times in clean water, then set them on the fire to stew in their own liquor; take the fish and beard them very clean, let the liquor settle, and strain it off, and take warm milk, and wash the fish very well; then take the liquor, some good gravy, and crumbs of bread; set it on the fire, and when the bread is a little stewed, take a quarter of a pound of butter, and roll it in fine flour to thicken it; then take an anchovy, a little mace and nutmeg; put in your fish and boil it half-a-dozen times, and serve it up."*

"*To stew Scallops.*—Boil them very well in salt and water; then take them out and stew them in a little of their liquor, a glass of white wine, and a little vinegar; add some grated breadcrumbs, and the yolks of two or three hard eggs minced small; stew all together till they are sufficiently done; then add a large spoonful of essence of anchovy, and a good piece of butter rolled in flour; or stew very gradually in a rich white sauce, with thick cream, until quite hot, without being allowed to boil, and serve with sippets.†

* From an old MS. B.—C. C. W.
† Murray's 'Modern Cookery Book,' p. 259.

"*To Cook Scallops.*—Clean them from the shell; take off the beards, as also the black marks they bear; then cut them into four pieces. Fry some breadcrumbs with butter, pepper and salt, to a light-brown colour. Then throw in your scallops, and fry all together for about three minutes and a half, taking care to shake the frying-pan all the time. Last of all, press them tight into shells or a dish, and brown them with a salamander, and send them to table."*

Fam. HALIOTIDÆ.
HALIOTIS.—EAR-SHELL.

HALIOTIS TUBERCULATA, Linnæus. *Ear-shell,* or *Venus's Ear.*—Shell ear-shaped; short flat spire, lateral, and nearly concealed; aperture wide; a longitudinal row of perforations on the left margin; the interior pearly and iridescent.

The *Ear-shell, Ormer,* or *Oreille de Mer,*† is said to take its place in the British fauna solely on account of its being found in the Channel Islands, where it is very abundant; but it is still more so on the coast of France, between St. Malo and Granville, and great quantities are brought from thence to the Jersey market, which is well stocked during the summer, and they are sold at the rate of sixpence a dozen. This celebrated shellfish has been praised by old authors as a most delicate morsel. One

---

* "A Man Cook." See 'Field,' February 20, 1864.

† In German it is the *Seeohr;* and Mr. Jeffreys gives the following names for the ear-shell: it is the *ormier,* or *si-ieu* (*six yeux*) of the French; the *patella reale* of the Sicilians, *lapa lurra* of the Portuguese, and *orecchiale* of the Italians.

writer speaks of the *ormer*, or *Auris marina*, " as a lump of white pulp, very sweet and luscious," and another, as quoted by Professor Ansted, in his 'Channel Islands,' mentions " a large shellfish, taken plentifully at low tides, called an ormond, that sticks to the rocks, whence we beat them off with a forck or iron hook. 'Tis much bigger than an oyster, and like that, good either fresh or pickled, but infinitely more pleasant to the gusto, so that an epicure would think his palate in paradice if he might but always gormondise on such delitious ambrosia." Athenæus also tells us that the ὦτια, or ears, are most nutritious when fried. Again, he says,* " But otaria (and they are produced in the island called Pharos, which is close to Alexandria) are more nutritious than any of the before-mentioned fish (speaking of cockles, sea-urchins, pinnas, etc.), but they are not easily secreted. But Antigonus, the Carystian, says this kind of oyster is called by the Æolians the ' Ear of Venus.'"

Captain Beechey, in his ' Voyage to the Pacific,' mentions the abundance of two species of Haliotis in the Bay of Monteroy, and that they are much sought after by the Indians, not only for food, but because the shells are used for ornaments, and the natives decorate their baskets with pieces of them. *Haliotis gigantea* is eaten by the Californian Indians, and the Chinese are very partial to Venus's-ears, which form part of a Chinese dinner, with sea-snails, sharks' fins, etc. The natives of New Zealand call *Haliotis iris*, the mutton fish.

In Guernsey, ear-shells are used by farmers to frighten away small birds from the standing corn—two or three of these shells being strung together and suspended by

* ' Deipnosophists,' vol. i. bk. iii. 35, p. 146.

a string from the end of a long stick, so as to make a clattering noise when moved by the wind.*

Haliotidæ in great quantities are brought to Birmingham from various parts of the world, for making mother-of-pearl ornaments, inlaying papier-mâché tables, etc., and also for making buttons. An instance has been known of a ship arriving at London from Panama, bringing more than two millions of pearl-shells for the English markets.

The wholesale price in the Channel Islands for shells of the first quality is £10 per ton, and by retail they are sold at 1*d.* per lb.

Mother-of-pearl, however, is not only made from the Haliotidæ, but the snail pearl-shell *Turbo cornutus*, the white pearl-shell, *Meleagrina margaritifera*, are also used in this manufacture. Curiously carved pearl-shells, the work of the monks at Bethlehem, are sold by them to pilgrims and others who visit the Holy Land, and Bruce states that mother-of-pearl inlaying was brought to great perfection at Jerusalem. The nacre was from the *Lulu el Berberi*, or Abyssinian oyster. Great quantities were brought daily from the Red Sea to Jerusalem, and crucifixes, wafer-boxes, and beads were made and sent to the Spanish dominions in the New World.†

In the days of luxury at Rome, the pannels in the golden house of Nero were of mother-of-pearl, enriched with gold and gems;‡ and dishes, bowls, and cups of pearl-shell, were greatly esteemed in the sixteenth and seventeenth centuries. Leland, in his 'Collectanea,'§ describes the christening of the child of the Lady Cicile,

* Jeffreys' 'British Conchology,' vol. i.; Introduction, lxix.
† Bruce's Travels; see appendix, vol. viii. p. 337, 338.
‡ 'Gems and Jewels,' p. 24. § Vol. ii. p. 691.

"wife to John, Erle of Este Frieseland, called the Marquis of Bawden, and sister to Eryke, King of Sweden, and the decorations of the chapel, &c. The christening took place at the 'Queene's Palleyes, Westminster,' 30th Sept., Anno 1565, and the chappell was hung with cloathe of gold. The communion table was richely furnished with plate and jewells, and amongst other ornaments were a 'Fountayne and Basen of mother-of-pearl, two shippes of mother-of-pearle, and another shipe of mother-of-pearl.'" Mr. G. R. Corner* mentions a very elegant cup in the possession of the Queen, made of staves of turbo-shell, mounted on a stem and foot of silver gilt. He also adds that the polished, but unmounted turbo, has been employed as a festive cup in Wales, to a comparatively late period.

We read also of a watch set in "mother-of-pearle, with three pendantes of gold, garnished with sparkes of rubies, and an opall in everie of them, and three small pearles pendent," which Lord Russell presented to Queen Elizabeth. The cathedral at Panama has two towers, with short steeples on them painted white, and these steeples are said by Mr. Elwes† to be faced with the large pearl oyster-shells; but they do not look well.

The scabbard of the sword of the Emperor Napoleon I., which he wore when First Consul, is of gold and mother-of-pearl; and mock pearls are now much used for jewellery made of the pearl-shell; the effect being nearly as good as real pearls, and far better than the most successful imitations in paste; and Theophilus, in his essay on various arts, speaks of "sea-shells which are cut into pieces, and filed as pearls, sufficiently useful

* 'Journal of Archæological Association,' vol. xiv. p. 344, 345.
† W.S.W., or a Voyage in that Direction to the West Indies.

upon gold."* Various kinds of shells are used for ornamental purposes, on account of their beautiful nacreous layer: *e.g.* a Mediterranean species of the little *Phasianella*, which is made into necklaces, earrings, etc., and known in England as "Venetian shells;" and in Paris I noticed some pretty bracelets, brooches, earrings, necklaces, and studs, made of the *Trigonia pectinata*, an Australian bivalve, so arranged as to show the bright pinkish-purple nacre inside the valves. The Miranhá Indians also wear on holidays a large button made of the pearly river-shell, in a slit, cut in the middle of each nostril;† and Sir Samuel Baker states that the women of the Shir tribe, living on the White Nile, make girdles and necklaces of small pieces of river mussel-shells, threaded upon the hair of the giraffe's tail, and that the effect is nearly the same as a string of mother-of-pearl buttons.‡

Through the kindness of Mr. Morton, of St. Clement's, Jersey, I am enabled to give the following recipe for cooking the sea-ear:—

"*To Dress them to Perfection.*—Take them out of the shells, and well scrub them; then let them simmer for two or three hours, until they are quite tender, after which they may be scalloped as an oyster, or put into the pan to brown with butter."

They require to be well beaten with a stick or hammer, to make them tender, if they are to be *fried*, and they are likewise sometimes pickled with vinegar.

* Theophilus, "qui et Rugerus," etc., translated by Robert Hendrie, chap. xcv. p. 391.
† 'Naturalist on the Amazon,' by H. Bates, vol. ii. p. 197.
‡ 'Albert Nyanza,' Baker, vol. i. p. 84.

Fam. PATELLIDÆ.
PATELLA.—LIMPET.

PATELLA VULGATA, Linnæus. *Limpet.*—Shell oval and conical in shape; apex central, or nearly so, strong, sometimes with ribs diverging from the apex to the margin, and sometimes quite smooth. Colours various, pale greyish-yellow or greenish-brown, inside generally showing the same colour through, and the markings of the ribs distinctly towards the margin; the inside of the apex an opaque bluish-white, and the whole slightly polished.

The common limpet is found distributed all round our coasts, where it is greatly valued as bait by fishermen, and Dr. Johnson calculates that in Berwick alone, there is an annual consumption of no fewer than 11,880,000 limpets for that purpose.* At low tide, limpets may be collected in great numbers from the rocks and boulders. Some are seen safely ensconced in holes or depressions made by means of the muscular action of their foot or disk, which is the width of the shell; others are seen creeping about in search of fresh resting-places, or food, with their tentacles slightly protruding beyond the shell, till alarmed by some touch or otherwise; and they adhere with wonderful strength to the rocks. Wordsworth says:—

> " And should the strongest arm endeavour
> The limpet from its rock to sever,
> 'Tis seen its loved support to clasp,
> With such tenacity of grasp,
> We wonder that such strength should dwell
> In such a small and simple shell."

On the Devonshire coast I have found them very

* Forbes and Hanley, Brit. Mollusca, vol. ii. p. 425.

large, and worn quite smooth; some specimens measuring as much as eight inches in circumference.

Limpets, a foot in diameter, are found on the Western coast of South America, and are used by the natives as basins.*

In many places limpets are used for food, especially on the Continent, where they are oftener eaten than the periwinkle. At Naples they make them into soup, and I am told it is an excellent dish. At Eastbourne we have often seen the Irish reapers come down to the shore and eat the limpets raw, which they had knocked off the rocks with their knives. The poorer classes at Eastbourne also eat them constantly; the children collecting them at low tide from the rocks. Mr. Patterson while residing, in 1837, near the town of Larne, co. Antrim, endeavoured to form some idea of the quantity of the common limpet taken from the rocks on that part of the coast, and used as food; and he had reason to believe that the weight of the boiled fish was above eleven tons. At Plymouth they gather great numbers of them, especially from the breakwater, as well as in the Isle of Man, where they are known by the name of "flitters;" and in Scotland the juice of these shellfishes is mixed with oatmeal. In the Feroe Isles they call them "flia;" and in 'Life in Normandy' (vol. i. p. 192) we are told "that limpets are constantly eaten by the poor; and that at Granville the children use a square-pointed knife, with a thick back, for getting them off the rocks; some having, in addition, small wooden hammers; others only a stone in their right hands. The edge of the knife was applied always on one side, and never on the top of the shell; a little sharp tap was

* Cuming, as quoted by Woodward, in 'Recent and Fossil Shells.'

given, either with the hammer or stone, and the fish fell at once." This reminds us of Hermippus, who says,—

"And beating down the limpets from the rocks,
They make a noise like castanets."*

The *Patellidæ* were also among the shellfish eaten by the ancients; Diphilus says they have a pleasant flavour, are easily digested, and when boiled are particularly nice.† It is a curious fact, and one which is puzzling to archæologists, that limpet shells should be found in such abundance in cromlechs, both in the Channel Islands and in Brittany, surrounding the remains of the dead, often covering the bones, skulls, etc., to the depth of two and three feet in thickness. Mr. F. C. Lukis, in the 'Journal of the Archæological Association' (vol. i. p. 28), mentions finding limpet shells, mixed with earth, round the bones, in the Cromlech du Tus, or de Hus, Guernsey. Again, in a "cromlech" in Jersey, discovered in April, 1848, Mr. Lukis adds that there is a difficulty in solving the great question—*why* such a mass of limpet shells should invariably accompany these abodes of the dead? They are found not only in the earliest deposits, but also amongst the more recent.‡

The term "Cromlech," as applied to the "*Cromlech du Tus*," is a local name, used in the Channel Islands for a subterranean chamber, lined with upright slabs, covered by a roof of one or more slabs of stone, with a long passage leading to it, formed in like manner of upright slabs covered by large lintels,—over which has been raised a tumulus of earth; while our term Cromlech is applied to those covered by one capstone only,

* Athenæus, Deipn., book xiv. 39.
† Athenæus, vol. i. book iii. p. 152.
‡ Journal of the Archæological Association, vol. iv. p. 336.

without any passage leading to them.* Those consisting of chambers and a long entrance passage, covered by slabs, within a large tumulus of earth, as at Wellow, near Stoney Littleton; at Rodmarton; at Uley; and at Nympsfield, are called *Tumps*. In speaking of Cromlechs, in the Channel Islands, I do not therefore allude to monuments such as we call Cromlechs; which last, though probably sepulchral, have not yet been found to contain interments.

Necklaces of limpets and other shells, strung together on fibre or sinews, are found in early British graves.

Limpet shells are also used for mortar.

In the island of Herm, near Guernsey, poultry are fed on *Patella vulgata*; but it is said that they will not touch *Patella athletica*, which is also considered too tough for bait.

Sea-birds feed on the *Patella*, and Mr. Gatcombe, in the 'Field,' August, 1863, mentions having once taken from the gullet of an oyster-catcher upwards of thirty limpets. He also adds an account of a curious occurrence which took place on the Plymouth breakwater some time ago:—One of the workmen employed on the breakwater observed a sandpiper fluttering in a peculiar manner, and discovered on approaching it, that it had been made prisoner by a limpet. It would appear that, in running about in search of food, the bird's toe had accidentally got under a limpet, which, suddenly closing to the rock, held it fast until the man came up, who with his knife removed the limpet, and released the bird.

The French call this shell *lépas, Patelle, Jambe, œil*

---

* See Sir G. Wilkinson, "British Remains of Dartmoor," vol. xviii. Journal Archæological Association, 1863.

de bouc,* and *Bernicle*; the Germans, *Schüsselmuschel, Napfmuschel*, or *Napfschnecke*; the Spaniards, *diampa*; the Portuguese, *lapa*; and the Italians, *lepade*; and in Cornwall, limpet shells are called *croyans*.

Limpet Soup.—Wash them, and free the shells from seaweed, etc.; put them into a saucepan and parboil them. Take them out of the shells; chop up some parsley, and put it, with a tablespoonful of oil, or an ounce of lard or butter, into a saucepan, and fry until it becomes brown. Add a pint of water, and, when boiling, throw in the limpets, with a teaspoonful of anchovy sauce, some pepper, and boil again for half an hour; or, if preferred, stew them before putting them into the soup.

To Dress Limpets.—Take those of a large size, and fry them with a little butter, pepper, and vinegar. The smaller ones are better boiled, and then eaten with vinegar and pepper.

Eastbourne method of Cooking Limpets.—Put them on the gridiron till all the water boils out of them, and then they are fit to eat.

Mr. Jeffreys speaks highly of *roasted limpets*, having tasted them in the island of Herm. The limpets were placed on the ground, and laid in their usual position, and cooked by being covered with a heap of straw, which had been set on fire, about twenty minutes before dinner.†

Limpet Sauce.—Choose clean-shelled limpets, not covered with barnacles, steep them in fresh water, and then beat them in a close covered saucepan until they part easily from the shells. They yield a rich brown liquor, in which, after being shelled, they may be stewed for half an hour. Thicken the liquor with butter and

* Jeffreys' Brit. Conchology, vol. iii. p. 241.
† Brit. Conchology, vol. iii. p. 239.

flour; strain and season with pepper, cayenne, and salt, and a slight flavouring of lemon-juice or vinegar. The limpets, being tough and indigestible, are not returned into the sauce.*

Fam. MURICIDÆ.
BUCCINUM.—WHELK.

BUCCINUM UNDATUM, Linnæus. *Whelk.*—Shell ovate, with eight whorls, more or less inflated, covered with transverse coarse striæ; waved or undulated obliquely, covered with a yellowish-brown epidermis; length about four inches. The aperture large, nearly half the length of the body whorl. Columella strong, pillar lip smooth, and bent back; interior white, very polished, sometimes lemon-colour, or orange; canal short; operculum of a reddish horn colour.

The shell of the common whelk, or buckie, the *Buccin ondé* of the French, varies very much in colour, being sometimes yellowish, without bands, and other specimens having chestnut spiral bands, or wavy blotches. White varieties are occasionally taken, and the shell figured being dredged up in deep water, has still the rough olivaceous-coloured epidermis on it. It is found often on the beach, and is a great enemy to other mollusks, boring holes in their shells, and sucking the juices of the fish within, by means of its spiny tongue. Dr. Harvey, in his 'Seaside Book,' says that "the proboscis of the whelk consists of two cylinders, one within the other, the outer of which serves for the attachment of the motor muscles, and the general protection of the organ;

* 'Practical Cookery,' p. 95, Hartlaw Reid.

while the inner, opening near the extremity with a longitudinal mouth, armed with two strong cartilaginous lips, encloses the tongue, and a great part of the œsophagus. The tongue is armed with short spines, and acting in concert with the hard lips, which can be opened or shut, or strongly pressed together, it forms a sort of rasp or auger, by which very hard substances are rapidly perforated; and then the tongue being protruded, the hooked spines with which it is armed, are admirably fitted for the collection of food."

Whelks are taken in great numbers in wicker baskets baited with offal, and Pliny describes the taking of the "purple fish" by a similar method, viz. in a kind of osier kipe, called *nassis*, baited with cockles.* Billingsgate Market is chiefly supplied from Harwich and Hull; and some of the steamers from the north bring six or seven tons at a time.† They are sold at 1*s*. 6*d*. to 2*s*. a measure; are in season from August to September, though they are really good to eat at any time. Children are frequently seen buying a saucer of whelks in London in the spring; and the shellfish shops near Billingsgate Market are well stocked with them. There are, as Woodward remarks, two different shellfish sold in London, under the name of "whelks," or "buckies," namely, the common *Buccinum undatum*, and the more prized *Fusus antiquus*. Whelks are very troublesome to the lobster-fishers, for they often devour the bait, and I have seen, at St. Margaret's-at-Cliffe, on the Kentish coast, the lobster-pots drawn up, one after the other, baitless, and full of these greedy mollusks; most trying to the poor fishermen, especially when bait was

---

* Pliny's Nat. Hist. vol. ii. bk. ix. p. 445.
† ' Curiosities of Food,' pp. 345–6.

scarce, and they had been obliged to walk some miles in the morning to purchase it.

On some parts of the coast the fishermen use the *Buccinum* for bait for the long-line fishing, and they know it by the following names, viz. the conch, buckie, whelk-tingle, or sting-winkle;* and at Youghal they call whelks "googawns," and "cuckoo shells."

In 'Popular History of the Mollusca,' by Miss Roberts, she mentions this species of shell being used in North Wales as trumpets by the farmers for calling their labourers; and shells of a similar kind are also used in Muscovy and Lithuania by the herdsmen for collecting their cattle, horses, mules, goats, and sheep. The Italian herdsmen use them also.

In some parts of Staffordshire the farmers call up their cattle by means of a horn or trumpet. In Tahiti shells were also used as trumpets,—a species of murex being the kind generally employed for that purpose. The largest shells were selected, sometimes a foot in length, and seven or eight inches in diameter at the mouth. A perforation, about an inch in diameter, was made near the apex of the shell, in which was inserted a bamboo cane, three feet in length, secured by being bound to the shell,—the aperture rendered air-tight by the outsides of it being cemented with a resinous gum from the breadfruit-tree. These shells were blown when any procession marched to the temple, and at other religious ceremonies; besides being used by the herald, and on board the native fleets. The sound is described as very loud, monotonous, and dismal.

We are told that in the island of Tanna, in the New Hebrides, shell trumpets are blown as signals to the

* A Book for the Seaside.

disease-makers, or sorcerers, to entreat them to stop plaguing their victims. "These disease-makers collected any *nahak*, or rubbish, that had belonged to any one, such as the skin of a banana he had eaten, wrapped it in a leaf like a cigar, and burnt it slowly at one end. As it burnt, the owner's illness increased; and if it was burnt to the end, he died: therefore, as soon as a man fell ill, feeling sure that some sorcerer was burning his rubbish, shell trumpets, which can be heard for miles, are blown as a signal to the sorcerers to stop, and wait for the presents which should be sent in the morning. When a disease-maker fell ill himself, he too believed that some one was burning his rubbish, and had his shells blown for mercy."*

The large chank-shell, *Turbinella rapa*, is a chief instrument of the Buddhists, who blow three times a day on this sacred shell, to summon believers to worship; and the same authority states that, according to the most ancient annals of the Cingalese, the chank-shell is sounded in one of the superior heavens of the demigods (similar to the conch-blowing tritons of Grecian mythology) in honour of Buddha, as often as the latter wanders abroad on the earth.† Sir J. E. Tennent mentions that this chank-shell is exported from Ceylon to India as a wind instrument, and also to be sawn into rings for anklets and bracelets; and also that a chank in which the whorls were reversed, and ran from right to left, instead of from left to right, was regarded with such reverence, that a specimen formerly sold for its weight in gold, but that now, one may be had for £4 or

---

* Turner, 'Polynesia,' as quoted in Taylor's History of Mankind, p. 128.

† 'Voyage of the Novara,' vol. i. p. 388.

£5. The Chinese also hold reversed chank-shells in special veneration, and give high prices for them. They are kept in the Pagodas by the priests and used on special occasions, and the consecrated oil is kept in one of these sinistrorsal *Turbinellidæ*, with which the Emperor is anointed at his coronation.* From the earliest ages, the Gulf of Manaar has been fished for chanks.

Dr. Potter, in his 'Archæologia Græca,' vol. ii., states that the ancient Greeks used shells as trumpets, as the Spaniards do at the present day; and that the first Grecian signals were lighted torches thrown from both armies by men who were priests of *Mars*, and that these signals being laid aside, shells of fishes succeeded, which were sounded in the manner of trumpets, which in those days were not invented. Hence Theognis's riddle may easily be interpreted:—

> "A *sea-inhabitant* with *living* mouth
> *Spoke* to me to go home, though *dead* it was."

Triton's shell-trumpet is famous in poetical story, whence Ovid, speaking of Neptune:—

> "Already *Triton* at his call appears
> Above the waves, a *Tyrian* robe he wears;
> And in his hand a crooked trumpet bears.
> The sov'reign bids him peaceful sounds inspire,
> And give the waves the signal to retire;
> His writhen shell he takes, whose narrow vent,
> Grows by degrees into a large extent."—*Dryden*.

And most of the poets mention this custom in their description of primitive wars.

The German name for the whelk is very appropriate, viz. *Trompetenschnecke*, or *Kinkhorn*. In Anglo-Saxon, whelk is *weolc*, but *weolc* is said to mean *that which*

---

* Lubbock's 'Prehistoric Times,' vol. i. p. 222.

gives the purple dye (therefore it would apply better to the dog-whelk, *Buccinum lapillus,* or *Purpura lapillus,* which yields a purple dye); thus, *embroidered with purple* is *weolc-basn-hewen;* scarlet dye is *weolc-read.* In 1684 *Purpura lapillus,* the dog-whelk, was employed for dyeing linen in Ireland; and Neumann says that the purple-fish was also found on the coasts of Ireland, and that some persons made considerable profit by marking linen with its juices.

The shell, which is very hard, is broken by a smart blow, taking care not to crush the body of the fish within. After picking off the broken pieces, there appears a white vein or reservoir, lying transversely in a little furrow near the head. This being carefully taken out, and characters drawn with it, or its viscid juice squeezed upon linen or silk, the part immediately acquires, on being exposed to the sun, a pale yellowish-green, which quickly deepens into an emerald green, then changes to a blue, and at last to a fine purplish-red. If the cloth be now washed with scalding water and soap, and laid again in the sun, the colour changes to a beautiful crimson, which suffers no further alteration from sun or air, soap, alum, alkaline leys, or any of the substances used for assaying the permanency of colours.

The juice of the purple-fish receives no colour itself, and communicates none to silk or linen, without exposure to the sun. It seems to be the light, and not the heat, of the sun, that calls forth the tincture; for when the cloth is covered with thin opake bodies, which transmit heat without light, no colour is produced, while transparent ones give no impediment to its production. The juice, itself, in close glass vessels, becomes presently purple in the sun.* Lister, in 1686, mentions the discovery of a shellfish, *Purpura Anglicana,* on the shores of the Severn, in which there is a vein, containing a juice, giving the delicate and durable tincture of the rich Tyrian purple. A writer in the 'Annual Register' for 1760 says that, being "at a gentleman's house in the west of Ireland," he "took particular notice of the gown of the lady of the house. It was a muslin flowered with the most beautiful violet colour. . . . She told me it was her own work, and took me to the seaside, where she gathered some little shells ; . . . beating them open and extracting the liquor with the point of a clean pen, she marked some spots directly before me." He adds :—"I suppose a hundred fishes would not produce a drop as large as a pea." Richard of Cirencester also mentions as a production of Britain, shells "from which is prepared a scarlet dye of the most beautiful hue, which never fades from the effect of sun or rain."

It is also stated in the 'Athenæum' of July 20, 1850, that the Nicaraguan Indians use a purple dye prepared from shellfish.

Pliny says that there are two kinds of fish that produce the purple-dye, the *Buccinum* and the *Purpura,* purple, or pelagia.† *Murex trunculus* is generally considered to have yielded it.

We all know the story of the discovery of the Porphyra shellfish, by the dog of a Tyrian nymph loved by Hercules; which having picked up some of these shells, and crushed them with its teeth, its mouth became stained with purple-dye. It is scarcely probable that it

---

* Neumann's 'Chemistry,' p. 510 ; the Memoirs of the French Academy for 1736. See Philosophical Transactions, No. 178.

† Pliny, Nat. Hist. vol. ii. bk. ix. chap. 67.

could crush the strong hard shells of the *Buccinum*, or *Murex*, but it might easily break the beautiful fragile shell of the *Helix ianthina,* which we know yields a purple juice; for though a fable, the above was intended to relate a possible event; and we are told by Sir Gardner Wilkinson that the *ianthina* is common on the coast about Tyre and Beyrout. And though so very small, being only the size of a small snail, three-quarters of an inch in diameter, the water becomes completely coloured all around it whenever it is alarmed, and throws out its purple liquid.*

Athenæus speaks of many different kinds of purple-fish, some of them of large size, like those which are found near Segeum and Lesteum; and some small, like those found in the Euripus, and around Caria. According to Pliny, the juice of the *Buccinum* was considered inferior by itself, but mixed with that of the *pelagia*, it blended well, and gave a bright lustre to the colour. The proper proportions for dyeing fifty pounds of wool, were 200 pounds of juice of the *Buccinum*, and 111 pounds of *pelagium*,† and this mixture produced a beautiful amethyst-colour. The Tyrian hue was given to wool by soaking it in the juice of the *pelagia*, while the mixture was in a raw state, and afterwards dipping it in the juice of the *Buccinum*. The best quality was of the colour of blood, of a blackish hue to the sight, but of a shining appearance when held up to the light.‡ The "conchyliated" colour comprehended a variety of shades, viz. that of the heliotropium, as well as one of a deeper colour; that of the mallow, inclining to a full purple, and

---

* See note, Rawlinson's 'Herodotus,' vol. ii. bk. iii. chap. 20, p. 415.
† *Pelagia* was the shellfish, and *pelagium* the juice, or colour, from it.
‡ Pliny, Nat. Hist. vol. ii. bk. ix. chap. 62 (38).

that of the late violet; this last being the most vivid of all the "conchyliated" tints.*

The best purple in Asia was that of Tyre, and the peculiar symbol of that city was the whelk, or *purpura*, and it appears on the Tyrian medals.† Strabo remarks that this city was rendered unpleasant as a place of residence, owing to the great number of its dyeing-works. In the days of Ezekiel, purple was imported by the Tyrians from the Peloponnesus, but they soon learned to extract the dye for themselves. A modern traveller, Mr. Wilde, observed at Tyre numerous round holes cut in the solid sandstone rock, in which shells seem to have been crushed. They were perfectly smooth on the inside, and many of them shaped like a modern iron pot, broad and flat at the bottom, and narrowing towards the top. Many of these were filled with a breccia of shells; and he supposes that all the shells were of one kind, probably *Murex trunculus*.‡

In Africa, the island of Meninx (now called Gerbee, in the Gulf of Cabes) was famed for its purple, as well as parts of Gætulia, that border on the ocean; and in Europe, the best came from Laconia.

Cornelius Nepos speaks of the *Tarentine* red; and Hardouin remarks that in his time were still to be seen the remains of the ancient dyeing-houses at Tarentum, and that vast heaps of the shells of the murex had been discovered.§

Aufrère, in 1789, describes a hill called Monte Tes-

* Pliny, vol. iv. bk. xxi. 22 (8). † Heraldry of Fish.
‡ W. Smith, Dict. of the Bible, vol. iii. p. 1581, article "Tyre." A friend of mine has also seen these holes, round, square, and oblong, 2 to 3 ft. deep, but doubts their containing a breccia of shells pounded up in ancient times.
§ Pliny, Nat. Hist., see note, vol. ii. bk. ix. ch. 63 (39).

taceo, behind the Alcantarine Convent, at Tarento, consisting chiefly of the shells of *Murex brandaris*, which were supposed to have produced the purple dye; and according to Dr. Bizio, the Tyrian purple was produced from this *Murex brandaris*, and the amethystine purple from *Murex trunculus*.‡ Romulus employed the purple-dye for the *trabea*. It was purple and white, something similar in cut to the *toga*, and was the royal robe worn by the early kings. Servius mentions two other kinds of *trabea* besides the one already described, one *wholly* of purple, which was sacred to the gods, and another of purple and saffron, which belonged to augurs. Julius Cæsar appears to have been the first of the Roman emperors who wore the *toga* entirely of purple.*

In 'Religious Ceremonies,' p. 309, we are told that the Pope celebrates Mass in Lent, Advent, and all eves on which fasting is required, in a purple robe.

Other shellfish produce purple dyes—amongst them, *Aplysia hybrida*, and I have dyed a piece of linen with the beautiful purple liquid which it emits, but it faded quickly.

Dr. Darwin mentions a large *Aplysia* which is common at the Cape de Verd Islands, five inches long, and of a dirty yellowish colour, veined with purple, which, when disturbed, emits a very fine purplish-red fluid which stains the water for a space of a foot round.

Scalaria communis yields a purple liquor destructible by acids, and *Planorbis corneus*, a purplish fluid, but it cannot be made of any use, though Lister tried several experiments with the vain hope of being able to fix it. In Spain, *Murex trunculus* is eaten, and *Pupura lapillus*

* Pliny, Nat. Hist. see note, vol. iv. bk. xxi. chap. 22.
† Cic. Philipp. ii. 34.

is said by M. Cailliaud to be used for food in the spring (after the fish have spawned) by some of the inhabitants of St. Michel-Chef-Chef, in the department of the Loire Inférieure.

The Almond Whelk, or Red Whelk as it is sometimes called, *Fusus antiquus*, is eaten at Liverpool; and great quantities are taken on the Cheshire coast. In Dublin, the fishermen use them principally for bait for the larger kind of fish, such as cod and ling, and only occasionally eat them boiled or pickled. The beautiful large white variety is dredged off the Irish coast. My largest specimen from Dublin measures $6\frac{1}{2}$ inches in length and $3\frac{1}{2}$ inches in breadth, and Mr. Jeffreys saw the shells used as lamps in the Shetland Isles by the northern fishermen. They are suspended from a nail in the wall or ceiling of the hut, by means of a piece of string, which is fastened round the shell in a triangular form. The inside is filled with fish-oil, and a wick of cotton or tow is put into the canal at the extremity of the mouth.*

In 'Antiquitates Culinariæ,' it is said that at the enthronization feast of William Warham, Archbishop of Canterbury, in 1504, 8000 whelks were supplied at five shillings a thousand, and they were served up as an accompaniment to sturgeon; and amongst the dishes forming part of the *second course*, we read of *Sturgeon in foyle with welkes*.

In heraldry we find whelks used, and the arms of Sir John Shelley, of Maresfield, in Sussex, are sable, a fess engrailed between three whelk-shells or. The Shellys of Lincolnshire bear, argent a chevron gules, between three whelks sable;† and the crest of Venables

---

* Jeffreys' 'British Conchology,' vol. i. p. lxviii. Introduction.
† Burke's 'General Armorie.'

of Cheshire, is a wyvern gules, issuing from a whelk-shell argent; and many other examples might be given.*

A *buccinum*, or *whelk*, with a figure rising out of it, or rather looking out of it, is sculptured on the font in St. Clement's Church, Sandwich.

Dublin Method of Cooking Whelks.—Cleanse them well, boil them till they can easily be taken from the shell, and then fry them with plenty of *fat* or butter, till they are brown.

Whelk Soup.—Take two onions and cut them into small dice, fry them in a stewpan with some butter; shake the pan well for a few minutes, add five heads of celery, two handfuls of spinach, two cabbage lettuces cut small, and some parsley. Shake the pan again, put in two quarts of water, some crusts of bread, a teaspoonful of pepper, and a blade or two of mace. Let this boil gently for an hour. Boil the whelks, take them out of their shells and fry them a good brown, then add them to the soup and let the whole boil a few minutes, then serve.†

Another Way of making Whelk Soup.—Wash the whelks well, boil them and pick them out of the shells. Put an ounce of butter or dripping, with some finely chopped parsley, an onion, a little pepper and salt, into a saucepan, and fry it until it becomes brown, adding a little flour. Then to this add a pint of water or a pint and a half of milk, and when it boils place in the whelks, and a teaspoonful of anchovy. Let it boil again for half an hour, then serve.

To dress Whelks.—Boil them till quite tender, then eat them with vinegar and pepper.

* Fairbairn's 'Crests of Great Britain.'   † Old Cookery Book.

Fam. LITTORINIDÆ.

LITTORINA.—PERIWINKLE.

LITTORINA LITTOREA, Linnæus. *Periwinkle.*—Shell spiral, solid; whorls six or seven in number, covered with longitudinal striæ; apex very pointed; aperture nearly round and large; pillar lip flat, broad, and white; outer lip sharp, sometimes white, and occasionally showing the colour of the exterior of the shell through. Interior of the shell a dark brown. Operculum dark horn-colour.

In Anglo-Saxon, the periwinkle is called *sea-snægl*, or sea-snail; in Ireland, the horse-winkle and *shellimidy forragy*, and at Belfast, *whelks*; in Cornwall, *gwean*; and in the north, *corvins*; and the French give it the name of *sabot*, or wooden shoe, as well as *vignot* or *vignette*, and *bigorneau*. In Brittany it is called, as elsewhere observed, *vrelin* or *brélin*;* and few persons who have paid a visit to the seaside can have failed to remark this common shell, which, at low tide, may be seen crawling over the tangled masses of seaweed. Many pleasant hours do children pass in gathering basketfuls of periwinkles, taking them home and boiling them, and enjoying a hearty meal, with the accompaniment of good thick slices of bread-and-butter. Periwinkles vary much in colour, some being of a dark olive-green, nearly black or of a pale greenish-white, like the specimen figured; and others red or rufous-brown, with narrow bands of smoke colour. Varieties of form also occur, and I procured from Exmouth two curious specimens, with the whorls angular and the edges sharp, instead of rounded.

* Jeffreys' Brit. Conch. vol. iii. p. 371.

Athenæus, in his 'Deipnosophists,' mentions several kinds of periwinkles. He says, "Of the periwinkle, the white are the most tender, and they have no disagreeable smell, . . . but of the black and red kinds the larger are exceedingly palatable, especially those that are caught in the spring. As a general rule, all of them are good for the stomach, and digestible when eaten with cinnamon and pepper."

There is a large consumption of these little mollusks in London; and Billingsgate market is supplied from various parts of the British coast; the largest supply is in May and June, and they sell at one shilling a measure. Mr. Patterson, of Belfast, states, in his 'Introduction to Zoology,' that quantities of periwinkles are annually shipped from Belfast for London, and in 1861 the amount was 3394 bags, each containing about three bushels, and weighing $3\frac{1}{2}$ cwt., so that the periwinkles exported in that year exceeded 10,000 bushels, and weighed nearly 600 tons.

In the Orkneys, at Stromness, I am told that they are collected in sacks, and sent south to the different markets.

Professor Simmonds states that the annual consumption of periwinkles in London has been estimated at 76,000 baskets, weighing 1900 tons, and valued at £15,000; further, that the inhabitants of Kerara, near Oban, gather them, and get sixpence a bushel for collecting them, and forward them from Oban to Glasgow, thence to Liverpool, *en route* for London. About 30 tons are sent up to London from Glasgow.

Mr. A. Morton tells me that in Jersey the market is supplied with periwinkles brought from Southampton, those found in the island being very small; and occasionally a few pints of the *Trochus* appear in the market and are sold as winkles.

The Chinese are very partial to *sea-snails,* and we read in a description given of a Chinese dinner that the second course consisted of a ragoût made of them. At Macao, these sea-snails are white, but at Ningpo they are green, viscous, and slippery, and by no means easy to pick up with chop-sticks. Their taste resembles the green fat of the turtle. It is curious that the most abundant shell found in the Scotch kjökkenmöddings is the periwinkle, and it is also met with in great numbers in the Danish shell-mounds.

Periwinkle Soup.—Take a pint and a half or a quart of periwinkles, wash them well, and boil them in a saucepan with a handful or two of salt, to enable you to pick out the fish easily. Put a little dripping or butter into a saucepan, with an onion or carrot, some chopped parsley, and a sprig of thyme, and fry until it becomes brown. Add a pint of water to this, and, as soon as it boils, put in the periwinkles (which have been previously picked out of their shells), with a little pepper and salt, and let the whole boil again for half an hour.

To boil Periwinkles.—It is only necessary to put them into a stewpan with as much water as will prevent the bottom from burning, as the liquor oozing from them will be sufficient for the purpose; when the shells open wide enough to extract the fish, they will be sufficiently done.*

Note.—It is necessary to throw into the stewpan a handful or two of salt with the periwinkles, otherwise half the fish could not be picked out. The "opening of the shell" refers, we conclude, to the falling out of the *operculum.*†

* Murray's 'Modern Cookery Book.' † M. S. L.

To stew Periwinkles.—Clean them and wash them from the sand in three or four waters, boil them and pick them out of their shells. To a pint of fish put half a pint of fish-stock, two ounces of butter, and some pepper and salt; add a spoonful of flour, stirred in gradually, and simmer over a slow fire until it is a proper thickness; add a large spoonful of essence of anchovy, and one of mushroom sauce.*

Fam. AVICULIDÆ.
PINNA.—SEA-WING.

PINNA PECTINATA, Linnæus. *Sea-wing.*—Shell wedge-shaped, gaping at one end and tapering to a point at the other, equivalve, horn-colour; hinge toothless, straight, and long; ligament linear, strong and elastic, and internal, sometimes smooth and at others with delicate ribs which radiate from the beaks, which are straight and pointed.

The *Pinna* is the largest of our British bivalves, and specimens are found twelve inches long and seven broad at the gaping end. Many pairs of this shell were found in the spring of 1862 on the beach at Dawlish, some of them with the fish still alive in them; but they were all small, the size of the one figured. Other localities mentioned by Forbes and Hanley are Salcomb Bay (where a bed of these shells was discovered by Montagu), Weymouth, and all the Dorset coast, Milford Haven, the Hebrides, Zetland, and in Ireland off the coasts of Londonderry, Antrim, Down, etc.; and at Youghal, where they are known by the name of "powder-horns," the

* Murray's 'Modern Cookery Book.'

fishermen bring in fine specimens from the "Nymph Bank." Mr. Jeffreys was informed by Mr. Spence Bate, that at Plymouth the trawlers call the *Pinnæ* "caperlongers," which word is supposed to be a corruption of *cappa lunga*,—the name they bear in the Mediterranean; and that the familiarity of Plymouth seamen with such Italian words is accounted for by so many of our men-of-war having been at Naples. They are also known in Italy by the following names:—*nacherone*, *madre-perna*, and *palostrega*. In France they call them *jambonneaux*; and in Germany, *Steckmuschel*.

The *Pinnæ* live in sand and mud, with the small end downwards, in an upright position, and attached by a very strong byssus of silky thread. A small species of crab lives frequently in the shell of the *Pinna*; and the following is a quaint description given by Pliny of the friendship of the *Pinna* and its little guest:—"The *pinna* is also of the tribe of shellfishes. It is always found in muddy places, but never without a companion, which they call *pinnoteres* or *pinnophylax*, and which is a little shrimp, or in some places a crab, a searcher for food. The *pinna* first gapes open, and, being destitute of sight, exposes its body within to various little fishes, which come leaping by close to it, and being unmolested, grow so bold as to skip into its shell and fill it full. The *pinnoteres*, waiting for the opportunity, gives notice to the *pinna* by a gentle pinch; upon which, shutting its mouth, it kills whatever is within its shell, and divides the spoil with its companion."*

Mr. Say† says that a small crab (a species of *Pinnotheres*) which lives in the shell of the common American

---

* Pliny, Nat. Hist. bk. ix. c. 42 (or 66, Tr. Bohn.)
† Journ. Acad. Sc. Phil. i. 68.

oyster (*Ostrea Virginica*) is much valued by oyster eaters in the United States; and that in opening a large quantity of oysters these little crabs are collected apart, and serve to gratify the palate of *gourmands*. They are only seven-twentieths of an inch long, by two-fifths wide.*

The byssus or silky thread of the *Pinna* is called by the Sicilian fishermen *lana penna*, and is manufactured into a silken fabric. It was known to the ancients, and called by them *pinna-wool*, and by the Tarentines *lana pesca*, or fish-wool. St. Basil, Bishop of Cæsarea, in Cappadocia, mentions it in one of his homilies, saying, "Whence had the *pinna* its gold-coloured wool,—that colour which is inimitable?"†

Gibbon states that the Romans called the *pinna* the "silkworm of the sea," and that a robe made from the silk was the gift of a Roman Emperor to one of the Satraps of Armenia.

In Aufrère's travels is a description of the mode of collecting these shellfish by the Neapolitans, and of the manufacture of different articles from the silk:—

"As soon as a *pinna* is discovered, an iron instrument, called *pernonico*, is slowly let down to the ground over the shell, which is then twisted round and drawn out. When the fishermen have got a sufficient number of them, the shell is opened, and the silk, called *lana penna*, is cut off the animal, and, after being twice washed in tepid water, once in soap and water, and twice again in tepid water, is spread upon a table, and suffered to become half dry in some cool and shady place. Whilst it is yet moist, it is softly rubbed and separated with the hand,

* Popular Hist. Brit. Crustacea, p. 54.
† Stolberg's Travels, vol. ii. p. 151, transl. by Thomas Holcroft.

and again spread upon the table to dry; and, when thoroughly dried, it is drawn through a wide comb, and afterwards through a narrow one. These combs are of bone, and resemble hair-combs. The silk thus combed belongs to the common sort, and is called *extradente;* but that which is destined for finer work is again drawn through iron combs or cards, called *scarde.* It is then spun with a distaff and spindle, two or three threads of it being mixed with one of silk; after which they knit, not only gloves, stockings, and waistcoats, but even whole garments of it. When the piece is finished, it is washed in clean water mixed with lemon-juice; after which it is gently beaten between the hands, and finally smoothed with a warm iron. The most beautiful are of a brown cinnamon, and glossy gold, colour. A pair of gloves made of the *pinna* silk may be seen in the British Museum; and in the International Exhibition some articles made of it were exhibited in the Italian Court, viz. a large shawl, gloves, and specimens of the thread in skeins.

As an article of food, the *pinna* is nearly as good as the scallop; and Plutarch tells us that Matron, the parodist, speaks of it as forming one of the dishes at an Attic banquet, saying—

"And pinnas sweet, and cockles fat were there,
Which the wave breeds beneath its weedy bed."

Indeed, if we may judge from the number of times Athenæus mentions it amongst the various eatable shellfishes, it formed a favourite article of food amongst the ancients, and was highly prized by them,—as it is at Naples in these days, where it is considered a *recherché* morsel, and too expensive for the poor people to indulge in. It is of greater value for its byssus than for the table.

Poli remarks that it rarely appears in the Neapolitan markets. He says that it is cooked at Naples with pepper, oil, and lemon-juice, and served with baked prunes.

The *pinna* may be cooked in the following manner:—

Pinna Soup.—Take five or six *pinnæ*, according to their size, and after they have been well washed, put them into a saucepan on a slow fire until the shells open; then take out the fish. Chop up some parsley very fine, and put it with a tablespoonful of oil or an ounce of butter, into a saucepan, and fry until it becomes brown. To this add a pint of water, and, when it boils, put in your fish, with a little salt and pepper,

Sometimes vermicelli is boiled with it, when more water must be added; or take a slice or two of bread nicely toasted, and, after cutting it up into small pieces, put it into the soup before it is served.

Fried Pinnæ like Cutlets.—Take half-a-dozen of these shellfish, and, after well washing them, place them in a saucepan over a slow fire until they open of their own accord; take out the fish from their shells, and place them on a dish, covering them well with flour or breadcrumbs. Put some oil or lard into a frying-pan, and, when it begins to boil, add your fish, and fry them of a bright yellow colour. The frying-pan should be gently shaken all the time, so that the fish may not adhere together, but be quite separate. Dried parsley may be added just before serving up.

Fam. VENERIDÆ.

TAPES.—PULLET.

Tapes pullastra, Linnæus. *Pullet or Cullyock.*—

Shell oblong, opaque; valves inequilateral, covered with concentric striæ, which become coarser and more wavy towards the extremities, and are crossed by longitudinal striæ; ligament external, long, horn-colour. Three teeth in each valve, erect, very narrow.

Though so common a species, the *Tapes* is not so generally eaten in England as abroad, though both this and *Tapes decussata* are eaten in Devonshire, Hampshire, and Sussex. They both inhabit muddy sand or gravel, and occasionally we find specimens of the former in holes which have been made by the *Pholas*, and deserted; and I have taken them out of holes in the rocks both at Tenby and Eastbourne, but rarely without some depression or distortion of the valves. But the *T. decussata* is more local than the *T. pullastra*. I had never found it in profusion till the spring of 1862, when, on visiting the sands near the mouth of the Exe, I noticed that at low-water mark the ground was covered with specimens of it, and also with *Scrobicularia piperita*, which is called by the Exmouth fishermen the "mud-hen;" but this latter is not used for food, as it has a hot biting taste.* It is a larger and more rugged shell than *Tapes pullastra*, though much resembling it, but it is not so convex, and differs from it in colour, being of a dirty white, with the bands, rays, or markings of a drab colour, sometimes of a purplish tinge; while *Tapes pullastra* is of a more yellowish-white, with zigzag markings of a rufous-brown, sometimes extending all over the shell, and at others only towards the extremities.

In the Northern Isles the pullet, or cullyock, is only used for bait.

* Jeffreys' Brit. Conchology, vol. ii. p. 446.

Tapes decussata is called in some parts of England "purr," and in Hampshire "butterfish." At Stubbington, near Tichfield, quantities are collected, and sold in the neighbourhood, at 5*d.* per quart, where they are considered richer and better than cockles. They are found at low tide not far from high-water mark, and their locality is easily detected by *two* holes in the sand or gravel (unlike the cockle, which makes but one) about an inch or so apart. They are easily dug up by means of an old knife. On warm still days they appear to rise more readily to the surface; but if cold or windy they burrow about two or three inches deep in the gravelly sand. Butterfish are considered very wholesome, and I was assured by the cockle-gatherers that they might be eaten with impunity at all times of the year, and never disagreed with people as the mussels and cockles occasionally do.

M. Gay says that at Toulon it is known by the name of *clouvisso,* and is a favourite dish in Continental seaports.* *Clovisse* is another name for it, and at Bordeaux it sells in the market from twenty to thirty centimes per hundred, and both it and *Tapes pullastra* are called *palourde* by the French. At Puerto de Santa Maria, in Spain, it is very highly prized, ana the Spaniards say "es buena" in speaking of it; and at Vigo thousands are gathered at every tide.

Other species of *Tapes* are eaten abroad, besides those already mentioned; and we may add another to our edible mollusks, viz. *Tapes Virginea,* which is distributed all round our coasts. It varies very much in colour, and you may gather a dozen or more specimens without finding two that resemble each other. The brightest I

* Jeffreys' Brit. Conch. vol. ii. p. 361.

ever found was near Dawlish; it was mauve colour, with white streaks. The largest are dredged at Tenby.

In Ireland, at Youghal, in Birterhury Bay, in Connemara, and in Bantry Bay, *Tapes aurea* is said to be eaten, but it is not a common species, though locally abundant; and in the spring, numbers are found in the Scilly Isles.

The Spaniards call the *Tapes, Almejas,* and, as I previously observed, prize it highly. At Cadiz, shellfish are considered good if people drink too much wine, and consequently they are often introduced at *festas*; and no food is considered by the Spaniards so nourishing as shellfish for those who work hard.

It is a rule at Spanish tables to hand round white wine with shellfish, though with other things they use any wine indiscriminately, and the wisdom of this custom is proved by experience. Indeed, serious illnesses are often caused by taking port wine with oysters, lobsters, etc.; the astringent qualities of port having the effect of hardening the shellfish, and sometimes producing violent indigestion.

The following recipes for cooking the *Tapes* are from Cadiz:—

Tapes Soup—Sopa de Almejas.—Wash the shells and put them into a saucepan with a little water, then put them on the fire for a few minutes to open them. Pick the fish out and put them into a clean saucepan, with an onion chopped very small, salt, pepper, and butter. Fry till they are of a good brown colour, then add water or broth, and boil till a strong soup is made. If preferred, fresh fish may be added when serving it.

Tapes decussata—Almejas blancas.—Wash them well, dry them, and place them in a saucepan or casserole in the oven, which must not be hot enough to burn them; when open, take them out of their shells, and place them on a very slow fire, with butter, parsley, and a little chopped onion; when tender, add a little flour, pepper, and half a glass of white wine. As soon as they are ready to serve, add the yolk of an egg well beaten, and the juice of a lemon.

Tapes, another way—Almejas cocidas.—Wash and open them as above, add butter and some chopped parsley, serve in their own liquor, with the juice of a lemon squeezed into it.

Tapes Ragoût—Almejas guisadas.—After having well washed the shells, put them into an earthen vessel, with a piece of butter; when open, pass the liquor that runs from them through a sieve, and take the fish out of the shells. Place the fish in the liquor, and add more butter, mixed with chopped parsley, pepper, and salt; moisten them with broth, white wine, or water; let them boil some minutes, and when ready to serve, add an egg well beaten, some lemon-juice or vinegar.

Tapes au naturel—Almejas al natural.—Prepare them as mentioned in the recipe above, then put the fish in a saucepan with their own liquor; add whole peppercorns and cook them over a very slow fire, shaking them about from time to time; then add lemon-juice and shake them again over the fire. Salt to your taste, and serve without any other sauce.

Tapes Sauce—Salsa de Almejas.—Scald the fish in boiling water to open their shells, but do not let them be heated more than necessary, clean them nicely, and mix them with a white sauce, acidulated with lemon-juice or vinegar; use with boiled or fried fish.

Potage of Oysters and Tapes—Menestra de Ostras y

Almejas.—Wash the shells and put them in hot water to open them. Take out the fish, and put them in a saucepan on the fire with a little water; chop two onions small and fry them in butter; while stirring them about dredge in slowly a little flour; add the oysters and Tapes, and the water in which they were boiled. Stir the whole for a few minutes over the fire, then add the yolk of an egg well beaten up. Fry slices of bread in butter, and place them at the bottom of the dish, pouring the potage over them; then serve.

Hampshire method of Cooking Tapes.—Wash the shells, then boil them for a few minutes, till the water is just on the eve of boiling over. If boiled with cockles, the "butterfish" must be placed in the saucepan a few seconds *before* the cockles. They are also very good eaten raw, like oysters.

VENUS VERRUCOSA, Linnæus. *Warty Venus.*—Shell opaque, very solid, inequilateral, covered with concentric edges which bend backwards, and towards the sides or ends become coarser, forming knots or tubercles. These ridges are divided by fine ribs or furrows, which radiate from the beaks, giving them a scalloped appearance. Umbones prominent, the beaks small and sharp, the lunule distinct and heart-shaped. Ligament rather long and narrow. Three teeth in each valve; the margins crenulated inside. Colour pale yellowish-brown.

This coarse, rough-looking shell is found on many parts of the coast of the English Channel, also in the Channel Islands, and in Ireland.

Mr. Hanley states that at Herm, near Guernsey, it is collected as an article of food from the small pools between the rocks at low water;* and Mr. Jeffreys says

* Forbes and Hanley, Brit. Mollusca, vol. i. p. 404.

that it is habitually eaten in County Clare, and that Weinkauff mentions its being sold in the market at Algiers.*

It is a common species on the south coast of Ireland, and Mr. Damon, of Weymouth, on visiting Kenmare, found that, owing to the great consumption of *Venus verrucosa* for food, the species was nearly exhausted. It is dug out of a sandbank at low spring tides, at Bantry. Dr. Paul Fischer observes that they are endeavouring to cultivate it on the coast of Provence, and that it ought to thrive well at Arcachon, as it is indigenous there.

The beautiful *Venus Chione*, or *Cytherea Chione*, may also be included in our list of "edible mollusks," though it is not sufficiently abundant to form any more than a rare and dainty dish with us, while, in the Mediterranean, it is a common species.

Poli, in his magnificent work, the 'Testacea utriusque Siciliæ' (to which more modern writers are so deeply indebted for their anatomical description of molluscous animals), mentioning this fish, under the names of *Venus Chione* and *Callista coccinea*, says it is most excellent, and that, though cooked in various ways (common to different shellfish), it is most delicious when simply cooked in oil or butter, with breadcrumbs, chopped parsley, and pepper and salt.

I was so fortunate as to procure a dozen beautiful specimens from Plymouth, the largest measuring $2\frac{1}{2}$ inches in length and $3\frac{1}{2}$ in breadth. The colour is a pinkish-brown, with rays of a darker shade; the epidermis is of a pale horn-colour, and transparent, showing the rays of the shell through, and is very glossy. The shell itself is solid and opaque. Specimens sent to

* Brit. Conch. vol. ii. p. 341.

me from the Mediterranean are the same as those found on our coasts, both as to size and colouring, but this is not the case with some of our other bivalves,—the *Isocardia cor*, for instance, attaining to a larger size with us than it does in the south of Europe.

Messrs. Forbes and Hanley give the following localities for *Cytherea Chione:*—Plymouth, Teignmouth, Mount's Bay (Jeffreys), and other parts of the coast of Cornwall.

To Cook Venus verrucosa.—Boil them, after first washing the shells well to free them from sand and mud, then fry them for a few minutes in a frying-pan, with a little butter or lard, adding pepper and salt according to taste.

Fam. TELLINIDÆ.
PSAMMOBIA.

PSAMMOBIA VESPERTINA, Chemnitz. *The Setting Sun.*—Shell of an oblong oval shape, equivalve, rather flattened, opaque; colour whitish, shading to a reddish-yellow at the beaks, with radiating rays of carmine and purplish-pink; epidermis of an olivaceous brown; ligament external, prominent, and of a horn-colour; beaks small; teeth, two in each valve; in the left valve, one tooth bifid.

The *Tellinidæ* are but rarely used for food in this country, though several kinds are used for that purpose abroad. With us, the *Psammobia vespertina* is stated by Mr. Jeffreys[*] to be eaten by the peasantry at Kenmare, and heaps of their shells may be seen round the huts.

[*] 'British Conchology,' vol. ii. p. 400.

Mr. Damon informed me that this pretty shell is dredged during the summer months in Bantry Bay, all the boats being then engaged in dredging sand and its contents, for the farmers, who use it as manure; and that out of the heaps of sand, etc., formed on the quay, the Psammobia and other shells are collected. It is only a locally abundant species, but is generally diffused. Large richly-coloured specimens are found in Birterbury Bay, Connemara, and Cornwall, Devon, Dorset, Northumberland, Pembrokeshire, Firth of Forth, and the Channel Isles, are a few of the localities given by Mr. Jeffreys.

Athenæus[*] states that Tellinidæ were very common at Canopus, and abound when the Nile begins to rise, and that the thinnest of these were the royal ones, which were digestible and light. For fish-sauces, both the Psammobia and the Donax, or Wedge-shell (which belongs to the Tellinidæ also), might be substituted instead of cockles; and, indeed, a species of the latter, which with us is very rare, viz. *Donax trunculus*, is sold in the markets at Naples, and is said by Poli to be one of the best kinds of shellfish, both for making sauce and for seasoning small rolls of bread. Mr. Jeffreys adds that, according to Philippi, it is still esteemed a delicacy in the south of Italy, and in Sicily is called *cozzola*.

It is also much eaten in Spain, and at Malaga is very common, and is cooked with rice.

On the French coast the Donax is very abundant, and is eaten by the poor people, but always cooked. In German it is called *stumpfmuschel*.

Spanish method of Making Fish Sauce.—Scald the fish in boiling water, sufficiently to make the shells

[*] Athen. Deipn. vol. i. bk. iii. c. 40.

open; but do not let them be heated more than necessary. Clean them nicely, and then mix them with a white sauce. To give a piquant flavour, add a little lemon-juice or vinegar.

Spanish way of Cooking all kinds of Shellfish.—Chop up a good quantity of garlic, onions, parsley, and red peppers (which last must be prepared by throwing them into boiling water, and rubbing off the skins with a dry cloth); scald the fish, and pick them out of their shells, then put all together in an olla (or round earthen-pot), with plenty of oil; fry them till a deep yellow. They may either be served thus, or when finished add some broth, boil it up, and serve it like thick soup.

The genuine Cadiz lovers of shellfish, however, consider that scalding the fish spoils it; they therefore prefer the *raw* fish being put at once into the oil and vegetables, and the dish is then sent to table with the shells in it.

Fam. MACTRIDÆ.

MACTRA.—TROUGH-SHELL.

MACTRA SOLIDA, Linnæus. *Trough Shell.*—Shell thick and opaque, of a yellowish-white colour, nearly equal-valved, covered at the sides with a brownish or drab-coloured epidermis; nearly triangular in form; ligament short and internal; beaks small; a V-shaped cardinal tooth in one valve, with a long lateral tooth on each side, and fitting in the opposite valve into deep grooves, with toothlike edges.

Of the Mactridæ, both *Mactra solida* and *Mactra stultorum* are sometimes eaten in England, but they are not considered *very* good, and are full of sand; though the former is eaten in Devonshire; and Mr. Dennis (as quoted by Mr. Jeffreys, in his 'British Conchology') says that the people of Newhaven, near Brighton, eat the *Mactra stultorum* also. It appears that in 1861, the steam dredging-machines were at work at the mouth of the harbour, and that they turned up *Mactra stultorum* in great numbers, so that the beach at high-water mark was covered by them.* They live buried in the sand not very far from low-water mark, and at no great depth from the surface. In Holland the shells of *Mactra stultorum* are used for making roads and footpaths; they are also burnt for lime, and the fish is eaten there. According to Poli, it is known in Italy by the name of *mezzana*, and at Naples, *gongola*. In German, *Mactræ* are called *trogmuscheln*. Our rare *Mactra glauca*, or *helvacea*, which is a much larger shell than either of the other kinds above-mentioned, and is at least three inches long by four broad, with longitudinal rays of a pale fawn, or a drab, colour, resembling slightly *Mactra stultorum*, is sold in the market at Brest; and at Granville is known by the name of *schias*. It is also found at Naples, and Poli speaks with evident satisfaction of its sweet and excellent flavour. It is taken in the Channel Islands, but we seldom find more than single valves upon our coast. Mr. King, of 190, Portland Road, sent me a magnificent specimen alive, some time since, which enabled me to examine the fish, and admire the beautiful colouring of its two short thick tubes, of a pale-yellow, shading to a rich orange; round the orifices were dark streaks of crimson, the cirri of the same colour as the tubes. The animal, however, varies in

---

* 'British Conchology,' vol. ii. p. 424.

colour; and another live specimen I received afterwards was not so bright.

Mactra subtruncata, or the lady-cockle, as it is called at Belfast, is said by Mr. Alder to be gathered at Lamlash Bay, and used as food for pigs, and in some parts it is used as bait by fishermen.

One other species of Mactra may be mentioned as edible, as it is eaten in the Channel Islands, and also in Spain, viz. *Lutraria elliptica*, very unlike the Mactridæ in appearance, and not tempting to look at. It is a broad flattish shell, about five inches long, and three in height, with a long tube, something resembling *Mya arenaria*. It lives in muddy estuaries, and at the mouths of rivers, buried to the depth of one and a half to two feet; and I have had some fine specimens from the mouth of the Towy, in Carmarthenshire.

Mr. Dennis* says the *Lutrariæ* are called *clumps* at Herm, and I am told by Mr. Morton, that the fishermen in Jersey know them by the name of *horse-shoes*. In cooking them, they are first boiled, then taken out of their shells and fried. *Lutraria oblonga*, which is a common species in some of the little muddy estuaries near Croisic and Piriac, on the coast of the Loire Inférieure, is said by M. Cailliaud to be very generally eaten, but it is a rare species with us, though it has been taken on the Devon, Cornwall, and Dorset coasts. Mactræ are also found in great quantities buried in the sandbanks on the coast of Chili.

" *To Dress Mactridæ.*—Boil them, and then eat them with pepper, salt, and vinegar."

* 'British Conchology,' vol. ii. p. 430.

Fam. MYADÆ.
MYA.—GAPER.

Mya truncata, Linnæus. *Gaper or Truncated Mya.*
—Shell equal-valved, suboval, gaping much at the small end, truncated and swollen at the other, covered with a pale greenish epidermis, which also continues over its long broad tube and mantles; valves wrinkled transversely; beaks depressed; umbones prominent, but unequal; a large spoon-shaped tooth in left valve, with a socket or hollow in the other; ligament internal.

Of the three species of *Myadæ* which inhabit our British seas, two of them are used for food, viz. *Mya truncata* (the one figured) and *Mya arenaria*, which last is much eaten at Naples. At Belfast this shell is called "cockle brillion,"* evidently the same name as that applied in Brittany to the winkle, viz. *vrélin* or *brélin*. They live buried in the sand or mud, in an upright position, at the mouths of rivers and estuaries near low-water mark, and at low tide their locality is known by the holes in the surface. It requires much labour and patient digging, sometimes to the depth of more than a foot, to procure a dish of these esculents, therefore they are not so common an article of food as others which are more easily gathered. In Orkney, *Mya truncata* is called *kunyu*, and is not only eaten, but is used as bait for cod-fishing. The Zetlanders call it *smurslin*, the Feroese, *smirslingur*. They eat it boiled. In German it is the *klaffmuschel*. On some parts of the Devonshire coast it is known as the spoon-shell, probably owing to the wide spoon-shaped tooth in the left valve. The length of a full-grown spe-

* Jeffreys, Brit. Conch. vol. iii. p. 65.

cimen is about 3 inches, by 2½ in breadth. *Mya arenaria* is larger than *M. truncata*, longer and more pointed at the gaping end, equally coarse and rugged in appearance, its colour varying according to the nature of the soil in which it buries itself. Montagu states that this species is eaten at Southampton, and called "old maid;"* but upon making inquiry there I cannot discover that they are now known by that name. In Chichester harbour and in Fareham Creek the poorer classes collect them for eating, and call them "pullers." At Youghal the name for them is "sugar-loons," and in Dublin "colliers," and at both places they are considered good bait, and fit to eat; but at Youghal they warn you to be careful to take off the skin which covers the outside of the shell and tube, as it is supposed to be poisonous. However, it is probably harmless, except in cases where it causes indigestion; but I believe that *Mya arenaria* has been known really to disagree with some people, and Miss Ball mentions a friend being very uncomfortable after eating one. The Hampshire people do not seem to have noticed this peculiarity. I cannot let this opportunity pass without expressing my sincere thanks to Miss Ball, for much valuable information, which she kindly sent to me from Ireland, respecting the various edible mollusks.

Mya arenaria (*Mye des sables*) may occasionally be seen exposed for sale in the market at Bordeaux.

Myadæ are widely distributed, and are not only food for man, but for the walrus and other northern animals, besides birds and fishes, which relish them greatly. Captain Tuckey, in his expedition to the river Zaire, or Congo, found that a species of *Mya* was much sought

* Forbes and Hanley, 'British Mollusca.'

after by the natives, and that three or four hundred canoes were met with near Draper's Islands, in which the people were busily engaged in dragging up these shellfish; having made temporary huts by bending and entwining living branches of trees, besides occupying caverns in the rocks with their families during the fishing season. The shells were opened, and the fish having been taken out was dried in the sun.

"*Youghal way of Cooking Sugar-loons.*—Boil them; take them out of the shell, and eat them with a little butter, taking care to cut off the outside skin."*

Hampshire Method of Cooking Myadæ.—Wash the shells well, then boil quickly for a few minutes; as soon as the shell opens, the fish is cooked. Do not them boil longer, as it makes them hard, and spoils the flavour. A little vinegar and pepper can be added as a relish.

Fam. PHOLADIDÆ.

PHOLAS.—PIDDOCK.

PHOLAS DACTYLUS, Linnæus. *Piddock.*—Shell equivalve, oblong-ovate, gaping chiefly anteriorly, inequilateral, thick, white exteriorly and inside polished; exterior covered with longitudinal furrows and concentric striæ, with sharp radiating spines; no hinge; beaks hidden with callosities; a flattened spoon-shaped tooth, which curves forward, in each valve; accessory valves four in number.

The perforating powers of the *Pholas* have for a length of time been a subject of discussion amongst naturalists,

* Miss Ball.

and appear likely to continue so. Some thought that by means of its foot it perforated the soft clay or stone which hardened round it; and a Dutch philosopher named Sellius, nearly 130 years ago, published an account of the *Teredo,* wherein he showed that its shell could not be the instrument of perforation; and asked how it was possible that the extremely tender shell of the young *Teredo* could make a hole in solid oak—a material ten times harder than itself. He also observed that the form of the tube is evidently not the result of an auger-like instrument, because it is broader at the bottom than at the top and sides.

Mr. Jeffreys, who quotes the above in his 'British Conchology,' agrees with Sellius that the foot or muscular disk and not the shell is " the sole instrument of perforation by the mollusca of stone, wood, and other substances, which is closely applied to the concave end of the hole, and is constantly supplied with moisture through the glandular tissues of the body." He adds, " By this simple, yet gradual process, the fibres of wood or grains of sand-stone may easily be detached or disintegrated, time and patience being allowed for the operation." Some naturalists believe that it is accomplished by means of an acid contained in the fish, by which it dissolves the calcareous rocks; while others maintain that the *Pholas* bores by using its shell as a rasp. This mechanical process is fully described by "Astur," who, from his own observations, has endeavoured to solve the problem, and who, to quote Mr. Buckland's words, is apparently the only person " *who has ever seen the Pholas at work.*" In the 'Field,' "Astur" published some time since an interesting description of the method by which this mollusk bores its habitation. He says, "Having procured several of these mollusks in pieces of timber, I extracted one and placed it loose in my aquarium, in the vague hope that it would perforate some sandstone on which I placed it. It possessed the powers of locomotion, but made no attempt to bore. I then cut a piece of wood from the timber in which it had been found, and placed the *Pholas* in a hole a little more than an inch deep. Its shell being about two inches long, this arrangement left about an inch and three-quarters exposed. After a short time, the animal attached its foot to the bottom of the hole, and commenced swaying itself from side to side, until the hole was sufficiently deep to allow it to proceed in the following manner. It inflated itself with water apparently to its fullest extent, raising its shell upwards from the hole; then, holding by its muscular foot, it drew its shell gradually downwards. This would have produced a perpendicular and very inefficient action, but for a wise provision of nature. The edges of the valves are not joined close together, but are connected by a membrane; and, instead of being joined at the hinge, like ordinary bivalves, they possess an extra plate attached to each valve of the shell, which is necessary for the following part of the operation. In the action of boring, this mollusk, having expanded itself with water, draws down its shell within the hole, gradually closing the lower anterior edges, until they almost touch. It then raises its shell upwards, gradually opening the lower anterior edges and closing the upper, thus boring both upwards and downwards. The spines (points) on the shells are placed in rows, like the teeth of a saw; those toward the lower part being sharp and pointed, whilst those above, being useless, are not renewed. So far for the operation of boring; but how to account for the holes fitting the shape of the animal inhabiting them? To this I fearlessly answer, that this is only the case when the *Pholas* is found in the rock which it entered when small. This mollusk evidently bores merely to protect its fragile shell, and not from any love of boring; and in this opinion I am borne out by my own specimens. The young *Pholas*, having found a substance suitable for a habitation, ceases to bore immediately that it has buried its shell below the surface of the rock, etc. It remains quiescent until its increased growth requires a renewal of its labours. It thus continues working deeper and deeper, and, should the substance fail or decay, it has no alternative but to bore through, and seek some fresh spot where it may find a more secure retreat."

At Amroth, near Tenby, is a submerged forest, the trees of which are completely perforated by the *Pholas;* and at spring-tides fine specimens may be collected. Montagu remarks that, whilst it is the general habit of shipworms (*Teredo navalis*, or *Teredo norvegica*) to bore parallel with the grain, the *Pholas* perforates the wood across the grain.*

Mr. J. G. Jeffreys mentions that Redi, in a letter to his friend Megalotti, describes the *Teredo* as being not only eatable, but excelling all shellfish, the oyster not excepted, in its exquisite flavour. Nardo also praises it, and wonders why the Venetians, who call it *bisse del legno*, do not eat it.†

The German name for the *Pholas* is very appropriate, viz., *die bohrmuschel, steinbohrer*, or pierce-stone; in

---

* Forbes and Hanley, Brit. Mollusca.
† Brit. Conch. vol. iii. p. 159.

France it is called *le dail commun, gite,* or *pitau;* and in Spain, *folado.*

An old fisherman told me that the *pudworm*, as he called it, was a very delicate fish; and he had often noticed on the Hampshire coast, that at low spring-tides, in the winter, when sharp frosts set in, and when that part of the shore, where these mollusks bury themselves, is left exposed by the tide, they are all killed. He was in the habit of collecting the *Pholas dactylus* as bait for white fish, digging them out of the clay or shale; and he added that if he kept them a day or so before using them, they changed colour, and shone like glowworms, even shone quite brightly in the water, some distance below the surface, when put on the hooks for bait.

This reminds me of the following quaint lines in Breton's 'Ourania,' quoted in Daniel's ' Rural Sports:'—

> "The glowworme shining in a frosty night
> Is an admirable thing in Shepheard's sight.
> Twentie of these wormes put in a small glasse,
> Stopped so close that no issue doe passe,
> Hang'd in a *Bow-net* and suncke to the ground
> Of a poole or lake, broad and profound;
> Will take such plentie of excellent fish
> As well may furnish an Emperor's dish."

The luminosity of the *Pholas* after death is referred to by Pliny, who says, " the *onyches* shine in the dark like fire, and in the mouth even while they are eaten;"* and, " that it is the property of the *dactylus* (a fish so called from its strong resemblance to the human nail) to shine brightly in the dark, when all other lights are removed, and the more moisture it has the brighter is the light emitted. In the mouth, even while they are

* Pliny, Nat. Hist. vol. ii. bk. ix. c. 51.

eaten, they give forth their light, and the same, too, when in the hands; the very drops, in fact, that fall from them on the ground, or on the clothes, are of the same luminous nature."*

Dr. Coldstream states that "the phosphorescent light of this mollusk is given out most strongly by the internal surfaces of the respiratory tubes, and that it is strongest in summer; and Professor John Müller has observed, that when *Pholades* are placed in a vacuum, the light disappears, but reappears on the admission of air; also, that when dried, they recover their luminous property on being rubbed or moistened."†

Many others have also made experiments with the *Pholas*, and have studied its phosphorescence, viz. Réaumur, Beccaria, Marsilius, Galeatus, and Montius. The two first mentioned endeavoured to render this "luminosity permanent, and the best result was obtained by placing the dead mollusk in honey, by which its property of emitting light lasted more than a year. Whenever it was plunged into warm water, the body of the *Pholas* gave as much light as ever."‡

Beccaria also found that a single *Pholas* rendered "seven ounces of milk so luminous that the faces of persons might be distinguished by it, and it looked as if transparent."§

Pholas dactylus, or the *long oyster*, as it is called at Weymouth, is not often eaten in England, but is generally used for bait. A Newhaven fisherman, however, told me they sometimes collect some for eating from

---

* Pliny, Nat. Hist. vol. ii. bk. ix. c. 87.
† Forbes and Hanley, vol. i. p. 107.
‡ 'Phosphorescence,' by T. L. Phipson, Ph.D., F.C.S., p. 105.
§ Ibid. p. 104.

the chalk boulders, etc., between Newhaven and Brighton; that they were much more pleasant to the taste than whelks; and they only scald or boil them for a few minutes.

In France, in the neighbourhood of Dieppe, a great many women and children, each provided with an iron pick, are employed in collecting them, either for sale in the market or for bait.*

I find from Mr. Morton that they are plentiful in Jersey, and are sold in the market boiled ready for eating. In Spain, the *Pholas* is considered as next best to oysters, and is sometimes eaten raw. All the *Pholades* are edible, and a large West Indian species, *Pholas costata*, is much prized, and is regularly sold in the markets of Havanna, as we are informed by Forbes and Hanley.

Athenæus recommends these shellfish, as they are very nutritious, but he adds that they have a disagreeable smell.†

The *Normandy method* of cooking the *Pholas* (*le dail commun*) is to dress them with herbs and breadcrumbs, or pickle them with vinegar.‡

Large quantities of this fish are sold in the markets of La Rochelle, and Captain Bedford says that the *Pholas crispata* is eaten by the poor of Oban.§

Fam. SEPIADÆ.
SEPIA.—CUTTLE.

Sepia officinalis, Linnæus.   *Common Cuttle-fish or*

* Jeffreys' Brit. Conch. vol. iii. p. 102.
† Deipnosophists, vol. i. bk. iii. c. 35, p. 146.
‡ 'Cottage Gardener,' vol. i. p. 382.
   Jeffreys' Brit. Conch. vol. iii. p. 114.

Scuttle.—The common cuttle-fish, *la sèche, seiche,* or *casseron,* of the French, is very generally eaten by our fishermen, and at Great Yarmouth they bring them in baskets to the houses for sale, recommending them as excellent and wholesome food. Cuttle-fish are often taken on the fishing lines, and will follow the bait to the surface, sucking it and holding fast by their long tentacles,* but we seldom find them alive on the shore, though their white *bones* are constantly picked up; and an immense number of these *bones* sometimes strew the beach from Beachy Head to Pevensey, while numbers float on the water. This was particularly the case there some years ago. It seemed as if there had been some epidemic amongst the cuttles which caused this great mortality, for certainly many basketfuls of bones might easily have been collected. They are not without their use; and at Liverpool, cuttle-bones are sold to the druggists for making tooth-powder, as much as 12 cwt. arriving at a time;† and Pliny says that the ashes of calcined shells of the *Sepia* were used for extracting pointed weapons which had pierced the flesh.‡

In Germany, it is called the *Blackfisch,* or *Tintenfisch.*

The animal is curious, very flat, with white stripes across its body, the groundwork being dark brown. The head is brown, as well as the arms, but the inside of the latter is white, and is furnished with four rows of suckers. Its two tentacular arms are very long, expanded broadly at the tips, and are also furnished with suckers. The beak is hard and black, shaped like that of a parrot.

---

* 'Sea Fish,' etc., by W. B. Lord.
† Phipson's 'Utilization of Minute Life.'
‡ Pliny, Nat. Hist. vol. vi. bk. xxxii. c. 43.

Cuttle-fishes are very common in the Mediterranean, and are highly prized by the Neapolitans. The modern Greeks also make them, and especially the *Octopodia*, a principal article of food; they dry them in great quantities, and store them away for use to be boiled or fried. Several kinds of *Cephalopoda* are eaten abroad. The *Octopus vulgaris* is eaten when young and small at Nice, where it is much more plentiful in the market than at Genoa; and if it weighs less than a pound, and is still tender, it is much esteemed. Those who purchase it generally hammer it well with a stick before cooking it; and it is also stated that the Greeks are careful to drag it for some time upon a stone, holding it by the opening in the body. The flesh is said to have a peculiar taste, consequently that of the cuttlefish and calamar (*loligo*) is preferred. At Naples, shellfish merchants of Sta. Lucia sell them ready cooked.*

These *Octopods*, called *Octopodia* by the modern Greeks, are regularly exposed for sale in the markets of Smyrna; as they are in the bazaars of India; and the North American Indians are also partial to them.

Plato, the comic writer, says:—

> "Good-sized polypus in season
> Should be boiled,—to roast them's treason,
> But if early, and not big,
> Roast them; boil'd ain't worth a fig."†

M. Verany gives the following description of it:— "The common *Poülp* [the *polpo* of the Italians] is scattered throughout the Mediterranean, and is found on the coast of the Atlantic at the Canaries. According to facts collected by M. D'Orbigny, it has

* See notes, 'Life in Normandy,' vol. i.
† Athenæus, Deipnosophists, vol. i. bk. i. c. 8, p. 8.

been found at Hayti, Cuba, Bahia, the Isle of France, the East Indies, and in the Red Sea. . . . This *Cephalopod* lives almost always amongst rocks, and generally hides itself in the holes and crevices, into which it penetrates with great ease, its body being very supple and elastic. It is in these recesses that he lies watching for the animals on which he lives; as soon as he perceives them, he cautiously leaves his den, darts like an arrow on his victim, which he wraps himself about, clasps in his serpent-like arms, and fixes, by means of his suckers. . . . Sometimes he places himself upon sandy ground at a short distance from the rocks, and is careful to construct a hiding-place. For this purpose he brings together, in the form of a circle, a quantity of pebbles, which he carries by fixing them on his arms by means of his suckers. Then, having formed a sort of crater, he ensconces himself in it, and there waits patiently for some fish or crab to pass, which he skilfully seizes." "The young *Poülps* in summer come to the pebbly shores, and they are sometimes met with in muddy places, from which they are taken by the trawl, together with numbers of Eledon (*Eledone cirrhosus*). They are usually fished for with a line without a hook, instead of which is substituted a piece of dog-fish, a bit of cuttle-fish, a white fish, a bone, a piece of suet, or some attractive substance weighted with a small stone. . . . They are also caught with a small olive-branch, fixed at the end of a rod, fitted with a hook, which is drawn backwards and forwards before the openings of the holes and crevices of the rocks."

M. Verany further states that the fishermen catch the large ones with the *leister*, or trident, and in summer the young Poülps are caught with a line weighted with lead, furnished with a cork fitted with several hooks, covered with pieces of scarlet cloth, twisted into thongs. He adds, that the largest Poülp he ever saw was about three yards long, and weighed nearly half a hundredweight. Poülps of thirty pounds weight are not rare at Nice, and those of twenty pounds are common.* In the Polynesian islands, the natives have a curious contrivance for catching cuttle-fish. It consists of a straight piece of hard wood, a foot long, round and polished, and not half an inch in diameter. Near one end of it, a number of beautiful pieces of the cowrie, or tiger-shell, are fastened one over another, like the scales of a fish, until it is nearly the size of a turkey's egg, and resembles the cowrie. It is suspended in a horizontal position by a strong line, and lowered by the fisherman from a small canoe, till it nearly reaches the bottom. The fisherman jerks the line to cause the shell to move, as if it were alive, and the jerking motion is called *tootoofe*, the name of the contrivance. The cuttle-fish, attracted by the cowries, darts out one of its arms, and then another, and so on, until it is quite fastened among the openings between the pieces of the cowrie, when it is drawn up into the canoe and secured.

Octopus vulgaris is rare on the British coast. I recollect that, about fifteen or sixteen years ago, one was found on the shore at Beachy Head, by two fishermen, who put it into a large bucket or tub, and took it round to most of the houses at Eastbourne for exhibition; and Mr. Gosse found one in 1860 on the beach at Babbicombe. Dr. Spence, of Lerwick, in 1862, sent an account to Dr. Allman, Professor of Natural History at

* See notes in 'Life in Normandy,' vol. i. pp. 293, 298.—D.D.

Edinburgh, of a huge cuttle-fish, which was thrown on shore somewhere on the Shetland Isles, its body measuring nine feet, and its arms sixteen feet in length.* Very large specimens are found in the Pacific, and also in the Indian seas, and the latter are said to seize canoes, and drag them down; and woe betide the unfortunate bather should he happen to be taken in the grasp of one of these monsters; and on the authority of Sir Grenville Temple, in Beale's 'History of the Sperm Whale,' an anecdote is given, showing what happened in the Mediterranean to a Sardinian captain, who was bathing at Jerbeh. He felt one of his feet in the grasp of one of one of these animals, and tried with his other foot to disengage himself, but his limb was immediately seized by another of the monster's arms. He then endeavoured with his hands to free himself, but these also in succession were firmly grasped by the polypus, and the poor man was shortly found drowned, with all his limbs firmly bound together by the twining arms of the fish; and it is extraordinary, that where this happened, the water was scarcely four feet deep. Frédol, in 'Le Monde de la Mer,' states that the famous diver Piscinola, who at the desire of the Emperor Frederick II., dived in the Straits of Messina, saw, with much alarm, enormous poülps attached to the rocks, their arms several yards long, quite capable of destroying a man.

Pliny gives a description† of the dangerous powers of the polypus for destroying a human being in the water; embracing his body, it counteracts his struggles, and draws him under with its feelers, and its numerous suckers. It is said that the fishermen at the present

---

* 'Life in Normandy,' notes.—D.D.

† Pliny, Nat. Hist. vol. ii. bk. ix. chap. 48, and note.

day, on the coast of Normandy, state that the polypus, which they call *chatrou*, is a most formidable enemy to swimmers and divers, for when it has embraced any of the limbs with its tentacles, it adheres with such tenacity, that it is quite impossible for a person to disengage himself, or to move any of his limbs.

A friend told me, that on his voyage to Ceylon, many years ago, he used to beguile the time by fishing, and once he caught a huge cephalopod. When it was hauled on board, it stuck and clung with such tenacity to the deck and ropes, that it could not be pulled off, and was at last cut to pieces with a hatchet.

M. Flourens communicated to the French Academy an account of an enormous specimen which was seen by Lieutenant Bouger, forty leagues north of Teneriffe. It appeared to be about twelve to fifteen metres in length (from thirty-one to forty-six feet), its body of a reddish colour, and shaped like a horn. The widest part was about two yards in diameter. M. Moquin-Tandon observes that the fishermen of the Canaries often met with these huge monsters, exceeding one or even two yards in length, but they were afraid to attack them.*

A sailor who had seen some very large ones at Bermuda, and had heard of people being attacked by them whilst bathing, told me that he had ever after felt shy of bathing in the sea, and that even the thoughts of them made him shudder.

The Norwegian Kraken, Kraxen, or Krabben, was held to belong to the Cephalopods, and old Eric Pontoppidan, a Norwegian bishop, describes it as "*an animal, the largest in creation, whose body rises above the surface of the water like a mountain, and its arms like

* 'Intellectual Observer,' vol. i. pp. 82–83.

the masts of ships;" and he adds, that a whole regiment of soldiers could easily go through their manœuvres on its back. The Bishop of Nidros is said to have discovered one of these gigantic krakens asleep in the sun, and believing it to be a large rock, raised an altar on its surface, and celebrated Mass. The kraken remained stationary during the ceremony, but the bishop had scarcely regained the shore, before the monster replunged into the deep.*

The *Hydra* of Lerna, destroyed by Hercules, was most certainly a *polypus,* or *sepia,* and, in *at least one* of the early representations of the subject, the animal is most correctly drawn as a cuttle-fish or polypus. Montfaucon represents the hydra as "a monster with several heads—some seven, others nine, and others fifty, but that it was not a dragon is evident, not only from the waves which are at its feet, but also from the form and capaciousness of its breast, and whole body; and, again, its connection with the ocean can be traced, in the *crab* being sent to its assistance by Juno, to bite Hercules in the heel, and when he crushed it, he overcame the *Hydra.*

Pliny mentions several kinds of polypi, one which he especially calls the *land polypus,* and states that it is larger than that of the sea; and Hardouin says it is the species found on the seashore, which more frequently comes on dry land than the other kinds.†

The polypus is recommended by Pliny for arresting hæmorrhage,—it is bruised and then applied; and he further adds, concerning it, that of itself it emits a sort of brine, and therefore needs none to be used when it is cooked; that it should be sliced with a reed, as it is

* Frédol, 'Le Monde de la Mer,' p. 310.
† Pliny, Nat. Hist. vol. ii. bk. ix. c. 46; see note.

spoilt if an iron knife is used, " becoming tainted thereby, owing to the antipathy which naturally exists between it and iron;" and * Dalechamps suggests, that this means, " it being the nature of its flesh to cling to the knife."

The flesh of the *loligo*, or *squid*, was highly esteemed by the ancients, and Ephippus recommends the eating of *squids* and *cuttle-fish* together.

> " And many polypi, with wondrous curls."
>
> Athen., Deipnosophists.

And Sotades, the comic poet, introduces a cook, speaking as follows:—

> "To these I added cuttlefish and squills;
> A fine dish is the squill when carefully cooked,
> But the rich cuttlefish is eaten plain ;
> (Though I did stuff them all with a rich forced-meat
> Of almost every kind of herb and flower)."
>
> Bk. vii. c. 41, Athen., Deipnosophists.

They are still exposed for sale in the bazaars and markets in India.

With us the *squid*, or *squill*, as it is sometimes called at Weymouth, is only used as bait. It is good for catching conger-eels and whiting-pout, also for cod-fishing; but it is also a great enemy to the fisherman, and on the French coast they say that the *calmar*, as they call it, often tears the fish from their hooks during the night, when they are fishing with lines. The inhabitants of the Basque provinces esteem *calmars* highly as food, and call them *chipirones*, and at Bayonne they are also known by the same name, as well as by that of *cornet* or *corniche*.

In Japan, *squids* are regularly collected for food, and Mr. Arthur Adams gives, in the ' Zoologist,' p. 7518, an

* Pliny, Nat. Hist. vol. vi. bk. xxxii. c. 42.

interesting account of the *squid fishery*, off Nisi-Bama, in the Oki Islands. On nearing the anchorage, on the 19th November, 1859, they were struck by the number of lights on the water, moving in all directions, and on inquiry they found that they were from fishing-boats on the look-out for *Ika-surame*, or *squids*. The lights were produced by kindling " birch-bark in small kinds of gratings, with long wooden handles,—machines known among seafaring men by the name of *devils*. The flame of the fires is very clear and vivid; and the devils, being held over the sides of the boats, attract the squids." They were a species of *Ommastrephes*, usually called by the fishermen the *flying-squids*, or *sea-arrows*, as they swim very rapidly over the surface of the water, in immense shoals. They were taken " by jigging." The " jig" is of iron, and consists of a long shank, surmounted by a circlet of small recurved hooks. These cuttles are favourite articles of food, both with Japanese and Chinese, and are carefully dried for the market, and sold in great quantities. Near Hakodadi, there is, we are told by Mr. Adams, a small fishing village exclusively devoted to the catching and curing of the *squid;* and many hundreds of thousands may be seen daily drying in the open air, all nicely cleaned; each kept flat by means of little bamboo stretchers, and suspended in regular rows on lines, which are raised on poles about six feet from the ground. The open spaces, and all the houses in the village, are filled with these squid-laden lines. Squids everywhere form a novel kind of screen.

Pliny speaks of the *springing loligo*, and Trebius Niger remarks that whenever it is seen darting above the surface of the water, it portends a change; and also that they sometimes dart above the surface in such vast numbers, as to sink the ships upon which they fall.*

Another of the Teuthidæ, which is rare on our coast, but is common in the Mediterranean, *Sepiola Rondeletii*, is eaten at Nice, and is called *supieta* or *sepiata*, and is said to be a very delicate morsel. The Italians call it *calamaretto* and *seppietta*, and quantities are consumed at Genoa, and at Leghorn.

Aristotle speaks of the *Teuthis*, which he says is a kind of cuttle-fish, but different from the *sepia*, and has ink of a pale colour. Alexis talks of cooking them thus:—

> "I took the *teuthides*, cut off their fins,
> Adding a little fat, I then did sprinkle
> Some thin shred herbs o'er all, for seasoning;"
> Bk. vii. c. 130, *Athen.*, *Deipnosophists*.

And Antiphanes, in his 'Female Fisher,' says (referring to the ink):—

> "Give me some cuttle-fish first. O Hercules,
> They've dirtied every place with ink; here take them
> And wash them clean."

According to Pliny, Anaxilaus states that the ink of the sepia is possessed of such remarkable potency that if it is put into a lamp, the light will become entirely changed, and all present will look as black as Ethiopians.†

The ink of the *cuttle*, or *sepia*, is dried, and imported from China to Liverpool, where it arrives either in cakes, or is there made into cakes called *sepia*, which is used in painting. Dr. Lankester, in his little work on 'Animal Products,' says that the cuttle-fish is very abundant in the Mediterranean, and that the ink-bag is carefully

* Pliny, Nat. Hist. vol. vi. bk. xxxii. c. 6. † Ibid. c. 52.

extracted, the liquid being poured out to allow of its drying as quickly as possible. It is then triturated with a little caustic soda or potash, and afterwards boiled with caustic lye for half an hour, when it is filtered, and the caustic liquid is then treated with an acid till it is neutralized. After standing, a precipitate falls, which is collected, washed with water, and finally dried by a gentle heat. This substance is the dark pigment used by artists under the name of *sepia*.

The polypus is the symbol of Messina, and, according to Montfaucon, is figured on a medal of that city, and a man's head on the reversed side.

In Spain the cuttle-fishes (*loligo ?*) "calamares" are eaten, and are either broiled on a gridiron, or stewed in red wine in an earthen jar; after which you may broil them if you like, or serve them in the wine, or stew them, adding, after they are tender, a little flour, and the yolk of an egg, well beaten, and this is considered the most wholesome way of dressing them.

Spanish Method of Stewing Cuttles.—Stew them over a *very slow* fire in oil or butter, and, before serving, add a little water, salt, breadcrumbs, saffron, and a *soupçon* of new honey or sugar.

"*Jersey Method of Cooking Cuttle-fish.*—Boil them for ten minutes, then take them out, and the skin will come off like a glove, leaving the fish like so many sticks of horseradish. Then boil them for an hour longer; take them out and cut them up, and fry them with onions. Some prefer slices of bacon fried with them, instead of onions, and served up with milk sauce."*

They are plentiful about October, and large ones are sold in the market at a penny each.

* Mr. A. Morton.

Italian Recipe.—Fry them in oil. They cook them thus at a small village on the Riviera, not far from Savona, and they taste like skate.

In France, *Octopus vulgaris* is highly prized for bait, and is also considered very good as food; and in 'Life in Normandy,' vol. i., is the following recipe for cooking it:—

"A dish of cuttlefish is divided in the centre by a slice of toast; on one side of the toast is a mass of cuttle-fish stewed with a white sauce; and on the other a pile of them beautifully fried, of a clear even colour, without the slightest appearance of grease. The flour of haricot-bean, very finely ground, and which is as good as breadcrumbs, is added."

"*Weymouth Recipe for Cooking common Cuttle or 'Scuttle.'*—Cut off the head and feelers, and take out the white bone; then boil for a short time till tender,— generally ten minutes or so will suffice. It is said to taste like lobster."

Alexis, in his 'Wicked Woman,' introduces a cook, saying:—

"Now these three cuttle-fish I have just bought
For one small drachma; and when I have cut off
Their feelers and their fins, I then shall boil them,
And cutting up the main part of their meat
Into small dice, and rubbing in some salt
(After the guests already are set down),
I then shall put them in the frying-pan,
And serve up hot towards the end of supper."*

Major Byng Hall mentions the cuttle-fish (*Calamares*) as one of the great treats of the natives of Madrid.†

* Athenæus, vol. ii. bk. vii. c. 124. † 'Queen's Messenger,' p. 341.

Fam. CIDARIDÆ.
ECHINUS.—SEA EGG.

ECHINUS SPHÆRA, Müller. *Common Sea-egg or Sea-urchin.*—A wish has been expressed that I should include the "sea-egg" in my 'Edible Mollusca,' but I scarcely feel justified in doing so, as it is *not a mollusk*, and has no other claim to appear on these pages further than from its being fit for food.

It belongs to another class of animals, the *Radiata* or *Echinodermata*, which includes the star-fishes and the *Holothuriadæ*. The *Radiata* are so called because all their parts radiate from a common centre.

Echinus sphæra is generally of a reddish colour, or purplish, and has white spines, in some tinged with purple.

Pliny states that the sea-urchin moves along by rolling like a ball, which is the reason that it is so often found with the prickles rubbed off; also "that these creatures foreknow the approach of a storm at sea, and that they take up little stones with which they cover themselves, as a sort of ballast; for they are very unwilling, by rolling along, to wear away their prickles. As soon as seafaring persons observe this, they at once moor their ship with several anchors."* By Aristotle it is called the "migratory fish." Professor Forbes, in his 'History of British Starfishes,' observes that "it is with their spines that the *Echini* move themselves, seize their prey, and bring it to their mouths by turning the rays of their lower edge in different directions. The mouth is generally turned to the ground, and the five teeth which project from it form part of a remarkable dental appa-

* Pliny, Nat. Hist. vol. ii. bk. ix. c. 51, p. 427.

ratus, known by the fanciful appellation of 'Aristotle's lantern.'"*

In heraldry we find, according to Mr. Moule, that the *Echinus* is borne, the arms of the Alstowne family being *gules, three sea-urchins in pale argent;* and those of Alstanton, *azure, three sea-urchins argent.* The shells of *Echinus sphærea*, the common sea-egg, are often used for making emery cushions, cases for yard measures, and other toys.

Pennant mentions sea-eggs being used for food in many parts of England; and Mrs. Gatty, in 'Old Folks from Home,' if I remember correctly, states that *Echinus lividus*, or "purple egg-urchin," is eaten on the west coast of Ireland. It is one of the burrowing species, and lives in holes formed by it in the rocks. Mr. W. Thompson informed Professor Forbes, that he had seen it in abundance in the South Isles of Arran. "It was always stationary, the hole in which it is found being cup-like, yet fitting so as not to impede the spines. Every one lived in a hole fitted to its own size, the little ones in little holes, and the large ones in large holes; and their purple spines and regular forms presented a most beautiful appearance, studding the bottoms of the grey limestone rocks' pools."†

At the Museum of the Jardin des Plantes, at Paris, I have seen specimens of this *Echinus* in a block of sandstone from the Baie de Douarnenez, in Finisterre; also, specimens of *Echinus perforans* in granite rock from the Bay of Croisic. How these animals bore into such hard substances is still a question; it is supposed by some that they first perforate with their teeth, and then soften the rocks by some secreted solvent. ‡

* Forbes's 'British Star-fishes,' p. 154.   † Ib. p. 170.   ‡ Ib. p. 171.

A friend of mine, who examined some of the holes, observes that they are evidently formed by the animal, and are lined with a smooth yellow substance which it deposits on the stone; that in limestone rocks the deposit is probably obtained from the stone itself by means of a solvent, but that in granite it may be derived from the lime held in solution in the sea water.

Echinus esculentus, the real *Oursin comestible,* is found in the Mediterranean, and also on the coast of Brittany; and I have seen specimens from the roadstead of Brest. Mr. R. Jones (as quoted by the Rev. J. Wood in his 'Natural History,' p. 722) gives a most amusing description of sea-egg fishing in the Bay of Naples, saying, " I had not swum very far from the beach before I found myself surrounded by some fifty or sixty human heads, the bodies belonging to which were invisible, and, interspersed among these, perhaps an equal number of pairs of feet sticking out of the water. As I approached the spot, the entire scene became sufficiently ludicrous and bewildering. Down went a head, up came a pair of heels; down went a pair of heels, up came a head; and as something like a hundred people were all diligently practising the same manœuvre, the strange vicissitude from heels to head, and head to heels, going on simultaneously, was rather a puzzling spectacle. On inquiry, it proved that these divers were engaged in fishing for sea-urchins, which are especially valuable just before they deposit their eggs,—the roe, as the aggregate egg-masses are termed, being large, and in as much repute as the 'soft roe' of the herring."

The Fuegian women dive to collect sea-eggs, both in winter and summer; and large sea-eggs are found in the Bay of Concepcion, which are highly esteemed by the Chilians, and eaten raw.

Echinidæ were also eaten by the ancients, and were said to be tender and full of pleasant juice, but apt to turn on the stomach; but they were considered good if eaten with sharp mead, parsley, and mint.*

Demetrius, the Scepsian, says that "a Lacedæmonian, once being invited to a banquet, when some sea-urchins were put before him on the table, took one, not knowing the proper manner in which it should be eaten, and not attending to those who were in the company to see how they ate it; and so he put it in his mouth with the skin or shell and all, and began to crush the sea-urchin with his teeth; and being exceedingly disgusted with what he was eating, and not perceiving how to get rid of the taste, he said, 'Oh, what nasty food! I will not now be so effeminate as to eject it, but I will never take it again.'"†

A friend of mine once tasted a sea-urchin raw, while she was travelling in the south of Europe, as it was highly recommended, and considered quite a delicate morsel; but she told me that it was very unpalatable, and rather bitter, and she had not the courage to swallow it like the Lacedæmonian.

At Marseilles, baskets are seen in the fish-market filled with the beautiful green sea-ribbon, *Zostera marina*, on which are placed sea-eggs, and generally one is broken to show the orange-coloured oval mass inside.‡

There are four species of *Echini* eaten, viz. *Echinus melo* (*l'oursin melon*), in Corsica and Algeria; *Echinus lividus* (*l'oursin livide*), at Naples; *Echinus esculentus* (*l'oursin commun*), in Provence; and *Echinus granulosus*.

* Athenæus, 'Deipnosophists,' vol. i. bk. iii. c. 41.
† Ibid. vol. i. bk. iii. c. 41, p. 152.
‡ 'Reise-Erinnerungen aus Spanien,' von E. A. Rossmässler.

Echinus esculentus is called in Feroese *eyilkier*.

They are usually eaten raw, like oysters, are cut into four quarters, and the flesh eaten with a spoon.*

To Cook Echini.—Boil them as you would boil eggs, and eat them with sippets of bread.

Generally considered in season in the autumn. The sea-egg becomes red like a crab when it is cooked, and is said to resemble it in flavour.

* 'La Vie et les Mœurs des Animaux,' par Louis Figuier.

LIST OF WORKS

REFERRED TO, OR CONSULTED, IN THE PREPARATION OF THIS WORK.

'A Book for the Seaside.'
Acton, Miss: 'Modern Cookery.'
Adams, Arthur: "Squid Fishing in Japan," 'Zoologist' for 1861.
Addison, J.: 'Remarks on Several Parts of Italy in the years 1701, 1702, 1703.'
Alcock, Sir J. Rutherford: 'The Capital of the Tycoon.'
Ansted and Latham: 'The Channel Islands.'
'Archæologia Cambrensis.'
'Archæological Association, Journal of the,' vols. i. ii. iv. xviii. and xx.
'Athenæum,' July 20, 1850.
Athenæus: 'The Deipnosophists, or the Banquet of the Learned.' Literally translated by C. D. Yonge, B.A. Bohn's Classical Library. 3 vols.
'Art Journal': "The Pilgrims of the Middle Ages." By Rev. L. E. Cutts. Vol. for 1861.
'Atlas Geographicus,' vol. i.
Audot, L. E.: 'La Cuisinière de la Campagne.'
Audot: 'Dictionnaire Géuéral de la Cuisine Française, Ancienne et Moderne.'

Aufrère, Anthony: Travels through various Provinces of the Kingdom of Naples in 1789 by Charles Ulysses; translated by.
Baines's 'Explorations in South-West Africa.'
Baird, W., M.D., F.L.S.: 'Cyclopædia of the Natural Sciences.'
Baker, Samuel White, M.A., F.R.G.S.: 'The Albert N'yanza, Great Basin of the Nile.' 2 vols.
Barrera, Madame de: 'Gems and Jewels.'
Bates, H. W.: 'Naturalist on the Amazon.' 2 vols.
Beckman's 'History of Inventions.'
Beechey's 'Voyage to the Pacific.' 2 parts.
Beltremieux, Édouard: 'Faune du Département de la Charente-Inférieure.'
Blackburn, Henry: 'Travelling in Spain in the Present Day.'
Blackwood's 'Edinburgh Magazine,' No. 561. July, 1862.
Blower, Ralph: 'A Rich Storehouse, or Treasurie of the Diseased.' 1607.
Bowles, W. L.: 'Poetical Works.' 2 vols.
'British Mollusca and their Shells.' By Messrs. Forbes and Hanley. 4 vols.
'British Monachism.' By Fosbroke.
'British Topography.' 2 vols.
Bruce's Travels. 7 vols.
Burke's 'General Armorie.'
Cailliaud, Frédéric: 'Catalogue des Radiaires, des Annélides, des Cirrhipèdes et des Mollusques marins, terrestres, et fluviatiles recueillis dans le département de la Loire-Inférieure.'
Camden's 'Britannia.'
Cautraine, F.: 'Malacologie Méditerranéenne et littorale.'
Chenu, Dr. J. C.: 'Manuel de Conchyliologie.' 2 vols.
'Chronicos de los Rel. Descalzos de S. Francisco.' By Juan Francisco de San Antonio. 1738.

Colborne, Robert: 'A Complete English Dispensatory,' etc., 1756.
Copley, Esther: 'Housekeeper's Guide.'
'Cottage Gardener.' Vol. i.
Cromwell's 'History of Colchester.' 2 vols.
Daniel's 'Rural Sports.' 4 vols.
'Dictionary of Practical Receipts.' By G. W. Francis.
'Dictionary of Greek and Roman Antiquities.' Edited by Dr. W. Smith.
Earl, G. W.: "On the Shell Mounds of the Malay Peninsula." 'Intellectual Observer,' vol. i.
Ébrard, Dr.: 'Des Escargots, au point de vue de l'Alimentation, de la Viticulture, et de l'Horticulture.'
Ellis, W.. 'Polynesian Researches.' 2 vols.
Elwes: 'W.S.W., a Voyage in that direction to the West Indies.'
'English Cookery Book.'
'Enquire Within upon Everything.'
Evelyn's 'Memoirs.' Edited by W. Brey, Esq.
Fairbairn's 'Crests of Great Britain and Ireland.' 2 vols.
'The Feroe Isles.' By the Rev. G. Landt.
'Field,' The.
Figuier, Louis: 'La Vie et les Mœurs des Animaux, Zoophytes et Mollusques.'
Fischer, Dr. Paul: 'Faune Conchyliologique marine du département de la Gironde,' etc.
Florez: 'Medallas de España.'
Forbes, E., M.W.S., For. Sec. B. S., etc.: 'A History of British Starfishes, and other Animals of the Class Echinodermata.'
Forbes, Edward: 'Malacologia Monensis.'
Forbes, James, F.B.S.: 'Oriental Memoirs.' 4 vols.
Francatelli's 'Cook's Guide.'
Frédol, Alfred: 'Le Monde de la Mer.'
'French Family Cook.'

Fuller, Thomas: 'Pharmacopœia Extemporanea.'
'Galignani's Messenger.'
Gatty, Mrs.: 'Old Folks from Home.'
Gell, Sir W.: 'Pompeiana.'
Gibbon's 'Decline and Fall of the Roman Empire.'
Gosse, Philip Henry: 'A Naturalist's Rambles on the Devonshire Coast.'
Gosse, Philip Henry: 'A Year at the Seashore.'
Gosse, Philip Henry: 'A Manual of Marine Zoology for the British Isles.' 2 parts.
Gosse, Philip Henry: 'The Aquarium.'
Grey's 'Australia.' 2 vols.
Gwillim's 'Heraldry.'
Hall: 'The Queen's Messenger.'
Harper, John, F.R.S.. 'Glimpses of Ocean Life.'
Harvey, W. H.: 'Seaside Book.'
Henderson, W.; 'Folklore of the Northern Counties of England.'
Hendrie, Robert: 'Theophili, qui et Rugerus,' etc. An essay upon various arts, etc. In 3 books.
Holcroft's 'Travels of Count Stolberg.' 2 vols.
Hone, William: 'Everyday Book.' 4 vols.
'Household Words.' Vol. iii.
Humphreys, H. Noel: 'The Coin Collector's Manual.' 2 vols. Bohn's Scientific Library.
'Illustrated London News.'
'Intellectual Observer.' Vols. i. ii. iii. and vii.
Jeffreys, John Gwyn, F.R.S., F.G.S.. 'British Conchology.' Vols. i. ii. and iii.
Jones, T. R.; 'The Aquarian Naturalist.'
Keogh, John: 'Zoologia Mediciualis Hibernica, or a Treatise of Birds, Beasts, Fishes, Reptiles, or Insects,' etc. 1739.
King, Rev. C. W.: 'Precious Stones, Gems, and Precious Metals.'
Kirby's "History of Animals," etc. 'Bridgewater Treatise.'

Kirby's 'Wonderful Museum.' Vol. ii.
'Knight's Encyclopædia.'
'Land and Water.'
Lankester, Dr.: 'Of the Uses of Animals, in Relation to the Industry of Man.' 2nd course.
Leland's 'Collectanea.' 6 vols.
Lord, John Keast : 'The Naturalist in British Columbia.' Vols. i. and ii.
Lord, W. B.: 'Sea-fish, and How to Catch them.'
Lubbock, Sir J.: 'Prehistoric Times.' 2 vols.
Lukis, F. C.: " Cromlech du Tus." 'Journal of the British Archæological Association,' vol. i. 1845-6.
Lukis, F. C.: " On the Sepulchral Character of Cromlechs in the Channel Islands." 'Journal of the British Archæological Association,' vol. iv. 1848-9.
Lyell, Sir C.: 'Antiquity of Man.'
Macgillivray, W.. 'Conchologist's Text Book.' Corrected and enlarged by.
M'Culloch's 'Commercial Dictionary.'
'Macmillan's Magazine,' No. 36, October, 1862. "The Fisher Folk of the Scottish East Coast."
'Magazine of Domestic Economy.'
'Maître Jacques.'
"Man Cook," A. See 'Field,' February 20, 1864.
'Meddygon Myddvai.' (Welsh MSS. Society, 1859.)
Miller, Hugh : 'Sketch Book of Popular Geology.'
Montfaucon, Antiquity explained and represented in Sculptures by the learned Father; translated into English by David Humphreys, M.A. 4 vols.
Moquin-Tandon, A.. 'Histoire Naturelle des Mollusques terrestres et fluviatiles de France.' 3 vols.
'Morning Post,' The.
Moule, Thomas : 'Heraldry of Fish.'
'Murray's Handbook to Kent and Sussex.'
Murray's 'Modern Domestic Cookery.'

"My Pearl-fishing Expedition." 'Household Words,' vol. iii.
'Natural History Review; a quarterly journal of Biological Science.' No. x., April, 1863.
Neumann's 'Chemistry.'
Nichols's 'Forty Years in America.' 2 vols.
Nicolas, Sir N. H.: 'History of the Royal Navy.' 2 vols.
'Normandy, Life in.' 2 vols.
'Notes and Memoranda,' "Gigantic Cephalopod." Vol. i., 'Intellectual Observer.'
'Novara, Voyage of the.' 2 vols.
'O'Brien's Adventures during the late War.' 2 vols.
'Old Cookery Book.'
'Oyster, the; Where, How, and When to Find, Breed Cook, and Eat it.'
Patterson's 'Introduction to Zoology.'
Phipson, Dr. T. L.: 'The Utilization of Minute Life.'
Phipson, Dr. T. L.: 'Phosphorescence.'
Picart, Bernard: 'Ceremonies and Religious Customs of the Various Nations.' 4 vols.
Pliny's 'Natural History.' Translated by the late John Bostock, M.D., F.R.S., and H. P. Riley, Esq., B.A. 6 vols. Bohn's Classical Library.
Poli: 'Testacea Utriusque Siciliæ.' 1795.
Pontoppidan, Erich: 'The Natural History of Norway.'
Potter, John: Archæologia Græca, or the Antiquities of Greece.' 2 vols.
Prescott, 'History of Ferdinand and Isabella.' Vol. i.
Quatrefages, A. de: 'Rambles of a Naturalist on the Coasts of France, Spain, and Sicily.' 2 vols.
Quincy, Dr. John: 'Pharmacopœia Officinalis.'
Rawlinson's Translation of the History of Herodotus. 4 vols.
Reeve, Lovell: 'British Land and Freshwater Mollusks.'
Reid, Hartlaw: "Practical Cookery," 'Handy Outlines of Useful Knowledge.'

Roberts, Mary: 'Popular History of Mollusca.'
Robinson, J. C.: Catalogue of the Special Exhibition of Works of Art, etc., on Loan at the South Kensington Museum. 1862.
Robinson's 'Essay towards a Natural History of Westmoreland and Cumberland.' 1709.
Robinson's 'Physical Geography of the Holy Land.'
Rosenhauer, W. von: 'Die Thiere Andalusiens.'
Rossmässler: 'Reise-Erinnerungen aus Spanien.' 2 parts.
Schilling, Samuel: 'Grundriss der Naturgeschichte des Thier- Pflanzen- und Mineralreich.'
Scott, Sir Walter: 'Marmion.'
'Shipwrecked Mariner:' "Visits to the Seacoasts." Vol. xii., 1865.
Shirley, Evelyn Philip: 'Noble and Gentle Men of England.'
Simmonds, Peter Lund, F.R.G.S.. 'Curiosities of Food.'
Smith, C. Roach: "Notes on some Leaden Coffins discovered at Colchester." 'Journal of the British Archæological Association,' 1846-7. Vol. ii.
Smith, C. Roach: "On Pilgrims' Signs, and Leaden Tokens." Vol. i., 'Journal of the British Archæological Association.' 1845-6.
Sowerby's 'Popular British Conchology.'
Sowerby's 'Conchological Manual.'
Soyer, A: 'Gastronomic Regenerator.'
Soyer's 'Ménagère.'
'Sporting Gazette,' December 24, 1864.
Street's 'Gothic Architecture in Spain.'
Strickland, Agnes: 'The Queens of Scotland, and English Princesses.' Vol. vi.
Swainson, W., F.R.S.: 'A Treatise on Malacology, or the Natural Classification of Shells and Shellfish.'
'Tabella Cibaria: The Bill of Fare, a Latin Poem. Implicitly translated,' etc.

Taylor's 'History of Mankind.'
Tennent, Sir J. E.: 'Natural History of Ceylon.'
The 'Times,' passim.
Troschel, Dr. Franz Hermann: 'Handbuch der Zoologie.'
Tupper, Martin: 'Proverbial Philosophy.'
'Voyages of the Adventure and Beagle.' 3 vols. and appendix. King, Fitzroy, and Darwin.
Walsh, J. H.. 'English Cookery Book.'
Warner, The Rev. Richard, of Sway, near Lymington, Hants: 'Antiquitates Culinariæ; or Curious Tracts relating to the Culinary Affairs of the Old English,' etc.
White: 'Popular British Crustacea.'
Wilkinson, Sir J. Gardner: 'Dalmatia and Montenegro.' 2 vols.
Wilkinson, Sir J. Gardner: "British Remains on Dartmoor." 'Journal of the British Archæological Association.' Vol. xviii. 1862.
Wilkinson, Sir J. Gardner: "Ostrea Virginica, a new British Oyster, at Tenby." See 'Zoologist,' 1865.
Williams, Rev. Charles: 'Silvershell, or the Adventures of an Oyster.'
Wilson, Dr. Daniel: 'Prehistoric Man.'
Wood, Rev. J.: 'Natural History.' 3 vols. (Fishes.)
Woodward: 'Manual of the Mollusca.'
Wordsworth's Poems.
Wright, Thomas: 'The Celt, the Roman, and the Saxon.'
'Zoologist,' The. 1860, 1861, and 1865.

INDEX.

Abbey seal, with figure of St. James, or St. Jacques de la Hovre, 104.
Acclimatization of Ostrea Virginica on the French coast, 81.
Action for trespass, 48.
African snails mentioned by Pliny, 11.
Allouret, or bird net, 45.
Almejas, or Tapes, 145.
Almejas al natural, 146.
Almejas blancas, 145.
Almejas cocidas, 146.
Almejas guisadas, 146.
American Clam acclimatized on the French coast, 102.
American oyster, Ostrea Virginica, 81.
Amethystine purple produced from Murex trunculus, 131.
Amount of oysters annually abstracted from the open sea, 73.
Amroth, submerged forest at, 159.
Ancient English cookery, 63.
Ancient Greeks used shells as trumpets, 127.
Anecdote of Dr. Black and Dr. Hutton, 18.
Animals adorned with pearls, 56.
Anklets and bracelets of chank shells, 126.
Annual Colchester oyster feast, 71.
Anodonta cygnea eaten in Leitrim, 62.
Anodontæ and Unionidæ used for bait, 62.
An other soueraigne Medecine for a Web in the eye, 4.
Antient cryes of London, 76.

Apicius discovers the art of preserving oysters fresh, 69.
Aplysia hybrida emits a purple liquid, 132.
Aplysia, large, common at the Cape de Verde Islands, 132.
Apple or vine snail, Helix pomatia, 1.
Aristotle's description of razor-shell, 39.
Aristotle and cartilaginous fish, 48.
Aristotle's Lantern, 176.
Arms of Buckenham Priory, 104.
Articles made of Pinna silk, 141.
Artificial oyster-beds of Great Britain, 70.
Ashes of calcined shells of Sepia for extracting weapons from wounds, 164.
Aspergille, or Helix aspersa, 14.
Athenæus and the Ephesian mussels, 52.
Athenæus recommends roasted Solens, 40.
Athenæus and the Tellinidæ, 150.
Aulo of the Romans, 39.
Aureille de mer, 113.
Auris marina, 113.
Australian freshwater mussels, 63.
Aviculidæ, 138.
Awabee or Awabi, 49.

Bags and pockets for mussels made of old nets, 46.
Bajaina, name for Helix aspersa at Grasse, 14.
Banarut, or Helix aspersa, 14.
Banded snail, Helix pisana, 2.
Baptismal shells, mentioned in a

list of Church ornaments in the fifteenth century, 112.
Baptismal shells, usually of silver gilt, but real scallop-shells used in some churches, 112.
Baptism, in private, a wooden shell used, 112.
Barrois, escargotière in, 12.
Baskets-full of razor-shells sold at Tenby, 41.
Beira, or great scallop (Pecten maximus), 101.
Belief in the power of the bones of St. James to work miracles nearly died out, 111.
Bérizon, or cockle, 27.
Bernicle, 121.
Bigorneau, 135.
Billingsgate Market supplied with mussels from Holland, etc., 47.
Birds feed on snails, 9.
Bishop Mayhew, 108.
Bisse del legno, 159.
Black cockle, 44.
Blackfish or Tintenfisch, 164.
Bohrmuschel or Steinbohrer, 159.
Bouchots, or artificial mussel beds, 46.
Bouger's huge cuttle-fish, 169.
Boyl-yas, or native sorcerer, 63.
Brading and its oysters, 73.
Breeding pearls, 30.
Bridge at Bideford, 48.
British localities for Solen marginatus, 39.
British oyster valued by the Romans, 68.
British specimen of Helix aperta, 15.
Brown oyster sauce, 86.
Bruvane, 27.
Bucarde, 27.
Buccinum, used for bait for longline fishing, 124.
Buccinum undatum, 123.
Buccinum, or whelk, carved on font in St. Clement's Church, Sandwich, 133.
Buccin ondé, 123.
Buckie, whelk-tingle or stingwinkle, 124.

Burran Bank oysters, 74.
Burton Bindona, oysters called, 74.
Butterfish, price of, 143.
Butterfish, or Purr, 143.
Byssus of mussels, 48.

Cæsar and the pearls of Great Britain, 55.
Cæsar, Julius, prohibits unmarried women to wear pearls and purple, 57.
Cæsar, Julius, first wore the toga entirely of purple, 132.
Cagouille, 14.
Calmar, 171.
Calamares eaten in Spain, 174.
Calamaretto, or Seppietta, 173.
Calcined mussel-shells make strong lime, 49.
Cancstrelli di mare, or Pecten varius, 99.
Caperlongers, 138.
Cappa di San Giacomo, 101.
Cappa Santa, 101.
Cappa tonda, 39.
Caracola, 19.
Caracola del huerta, 19.
Caracola del mar, 19.
Caracola del rio, 19.
Caracoleros, 19.
Caracoles con Perejil, 23.
Caragoou, 14.
Caraguolo, 14.
Cardiadæ, 27.
Cardium aculeatum found on the Devonshire coast, 38.
Cardium edule, 27.
Cardium rusticum, 36.
Cardium rusticum or tuberculatum, found at Paignton and Dawlish, 37.
Cardium rusticum, its leaping powers described by Mr. Gosse, 37.
Carlingford oysters, 74.
Carrickfergus oysters, large size, 74.
Cathedral at Panama, the steeples faced with pearl oyster shells, 116.
Catherine de Medicis, 58.

INDEX. 191

Cats made of the shells of Helix aspersa, 22.
Cawdel of Muskels, 64.
Cephalopoda, 165.
Cephalopods, large, at Bermuda, 169.
Cephalopoda, large, caught on a voyage to Ceylon, 169.
Ceylon pearl-fishery suffered from skate destroying the oysters, 61.
Chank shell used by the Buddhists, 126.
Chank shells exported to India from Ceylon, 126.
Chank shells reversed, prized by the Chinese, 126.
Chank fishery, 126.
Charity oysters, 61.
Charron, 45.
Chatrou, 169.
Cheyney, rock-oyster fisheries, 72.
Chilian method of cooking shell-fish, 67.
Chinese dinner, 136.
Chiocciola, Helix vermiculata, 17.
Chipirones, 171.
Christening of the child of Lady Cicile, wife of Erle of Frieseland, 115.
Cidaridæ, 176.
Cinque-Cento ornaments, 57.
Clams, several species of shells called, 101.
Clams acclimatized on the French coast, 102.
Clams strung like dried apples and smoked for winter use, 102.
Clams salted and preserved, 102.
Cleopatra and the pearl, 56.
Clodius Æsopus gives pearls to his guests to swallow, 56.
Closheens, 99.
Clouvisso, 144.
Clovisse, price of, at Bordeaux, 144.
Clumps, or horse-shoes, 153.
Cocciola, 39.
Cocciola zigga, 43.
Cochlea, 31.
Cochlear, cochleare, or cochlearium, 31.

Cockenzie fishermen, 76.
Cockles, 27.
Cockles live in sand, 27.
Cockles boiled in milk, 27.
Cockle brillion, 154.
"Cockle" applied to any shell, 29.
Cockle, or escallop, 29.
Cockles fried, 36.
Cockle-gardens, 28.
Cockle gatherers, dress of, 28.
Cockles, mussels, and oysters found on the sites of Roman stations, 34.
Cockle pies, 36.
Cockle porridge, Soyer's, 35.
Cockle, red-nosed, found at Paignton, 37.
Cockle, red-nosed, cooked Paignton method, 38.
Cockles said to yield a dye, 31.
Cockles sent to London from Gower, 28.
Cockles at Seville, 36.
Cockle-shells in an old British camp in Gower, 34.
Cockle-shell figured on coins, 28.
Cockle-shells prized by the Damaras, 28.
Cockle-shell used in heraldry, 28.
Cockle-shells used as leads on fishing nets, 28.
Cockle said to be used for skimming milk, 29.
Cockle sauce, 35.
Cockles scalloped, 35.
Cockled snails, 31.
Cockle soup, Francatelli's, 34.
Cockle soup, 38.
Cockles stewed in oil at Madrid, 39.
Cockles, to stew, 35.
Cœur-de-bœuf, or Heart-shell, 43.
Cog variously written, viz. kogge, gogga, kogh, cocka, coqua, etc., 34.
Cogs, vessels called, 34.
Colchester and its oysters, 71.
Cold weather injurious to the spat of the oyster, 70.
Colimaçon, or Helix aspersa, 14.
Colourists' shells, 49.
Concha di San Dialogo, 101.

Conchyliated colour comprehended various shades of purple, 130.
Consumption of the Apple or Vine Snail in Paris, 17.
Consumption of oysters in America, 81.
Contar, 14.
Coque, 27.
Coquilles de St. Jacques, 101.
Cornaillot, or Perceur, 70.
Cornets, or Corniches, 171.
Corvins, or periwinkles, 135.
Cotton wool injurious to pearls, 30.
Coutoye, 41.
Cozza di San Giacomo, 101.
Crab found in Ostrea Virginica, 139.
Crogans, Cornish name for limpet-shells, 121.
Cromlech, term, 120.
Cromlech du Tus, 120.
Crotalia or castanet pendants, ear-rings so called, 56.
Cullis of mussels, 67.
Cultivation of oysters on the western coast of France, 77.
Cup made of staves of turbo shell, 116.
Cups and dishes of pilgrims, 103.
Curried oyster atlets, 87.
Curried oysters, 88.
Cuttle-fish, or scuttle, 163.
Cuttle-fish, description of, 164.
Cuttle bones, 164.
Cuttle bones brought to Liverpool, 164.
Cuttles on the Sussex coast, 164.
Cuttles very large in the Pacific, 168.
Cuttle drowns a Sardinian captain, 168.
Cuttle-fish eaten by the modern Greeks, 165.
Cuttle-fish taken on fishing lines, 164.
Cuttle-fish, Italian recipe for cooking, 175.
Cuttle-fish, Jersey method of cooking, 174.
Cuttles, Spanish method of stewing, 174.

Cuttles or scuttle, Weymouth recipe for cooking, 175.
Cyprina Islandica, called the clam in the Shetland Isles, 101.
Cyprinidæ, 42.
Cytherea Chione or Venus Chione, 148.
Cytherea Chione, specimens of, from Plymouth, 148.

Dail, gite or pitau, 161.
Danes in the eighteenth century eat snails, 12.
Danish Kjökkenmöddings, 32.
Danish Kjökkenmöddings, oyster-shells in, 83.
Dartmouth oyster bed, 73.
Decoction of snails, Decoctum Limacum, 4.
Decoction of snails against consumptions (Decoctum Antiphthisicum), 5.
Demoiselles, 16.
Diámpa, 121.
Dijon method of keeping snails, 13.
Dijon way of cooking snails, 25.
Discovery of the ashes of St. James of Compostella, 107.
Distorted and deformed pearl mussel shells often contain pearls, 54.
Dog of Tyrian nymph, 129.
Donax eaten on the French coast, 150.
Donax cooked with rice at Malaga, 150.
Donax, called cozzola in Sicily, 150.
Donax and Psammobia used for making sauces instead of cockles, 150.
Donax trunculus sold in the markets at Naples, 150.
Dredgers of Whitstable, 72.
Dreissena polymorpha, 62.
Dress of Anne of Cleves, 57.
Ducks fed on snails, 21.
Duke of Bedford, arms of the, 104.
Dutch oysters, 89.

Ear-shell, Haliotis tuberculata, 113.
Ear-shells used in Guernsey by farmers to frighten birds from the corn, 114
Ear of Venus, 114.
Eastbourne oyster beds, 73.
Echinidæ eaten by the ancients, 179.
Echini, to cook, 180.
Echini best in autumn, 180.
Echini eaten raw like oysters, 180.
Echini move by means of their spines, 176.
Echinus in holes of rocks, granite, sandstone, and limestone, 177.
Echinus esculentus, 179.
Echinus esculentus the real oursin comestible, 178.
Echinus granulosus, 179.
Echinus in heraldry, 177.
Echinus lividus, or purple egg urchin, eaten on the west coast of Ireland, 177.
Echinus lividus eaten at Naples, 179.
Echinus melo, 179.
Echinus sphæra, 176.
Echinus sphæra, shells of, 177.
Eledone cirrhosus, 166.
Elenchi, long pearl-shaped pearls called, 56.
Enemies of the oyster, 70.
Enthronization feast of William Warham, 133.
Escallop in heraldry borne not only as a pilgrim's badge, 105.
Escallop shell, crest of Bower and of Bullingham, 104.
Escargotières, or snail gardens, 12.
Escargots, 14.
Escourgol, 14.
Esnandes, 45.
Experiments by M. Cuzent on green oysters, 79.
Export of snails from Saintonge and Aunis to Senegal and the Antilles, 15.
Extracting copper from oysters, 51.
Extravagance in jewellery from the 12th to 16th centuries, 57.

Eyilkier, 180.

Falmouth oysters sent to Marennes, 79.
Famine of 1816 and 1817, 15.
Fish and oyster culture company, 75.
Fishing for mussels in Bay of Concepcion, 50.
Flia, 119.
Flitters, 119.
Folado, 161.
Foreign pearls, 54.
Fortunes predicted by snails, 21.
Fountain of shells, 111.
French mussel breeders, 47.
French names for limpets, 121.
French names for scallops, 101.
Fried oysters another way, 89.
Frills or queens, 99.
Fuegian women dive for sea-eggs, 178.
Fusus antiquus, red or almond whelk, 132.
Fusus antiquus used as food, 132.
Fusus antiquus, shells of, used as lamps, 133.
Fusus antiquus sold in London under the name of whelk, 124.
Fusus antiquus, white variety, 133.

Gambling by means of snail races, 21.
Gaper, or Mya, 153.
Garden snail, Helix aspersa, 1.
Garden walks made of cockle shells, 28.
Gathering cry of pilgrims, 108.
Glamorganshire way of pickling oysters, 94.
Glow-worm, lines on a, 161.
Gofiche, or scallop, 101.
Goggle, or whelk, 29.
Gongola, or Mactra, 152.
Googawns and cuckoo-shells, 124.
Gower, a Flemish colony, 28.
Gower people live on cockles, 28.
Gower method of cooking cockles, 36.
Gower recipe for oyster soup, 85.
Grand'-pèlerine, 101.

o

Granville fisheries, 77.
Great drought in Ireland in 1792 or 1793, 52.
Green oysters in France, 78.
Grilled oysters, 90.
Grosille, 101.
Guisado de Caracoles, 24.
Gwean, or periwinkle, 135.
Gwillim's Heraldry, 22.

Habits of snails studied by the ancients, 22.
Haliotidæ, 113.
Haliotidæ brought to Birmingham, 114.
Haliotis gigantea eaten by the Californian Indians, 114.
Haliotis Iris, or mutton-fish, 114.
Haliotis tuberculata, 113.
Hardships of pearl divers, 60.
Hélices Vigneronnes method of cooking, 26.
Helicidæ in the markets in Murcia and Valencia, 19.
Helicidæ as Lenten fare, 13.
Helix aperta, 15.
Helix arbustorum, 10.
Helix aspersa, garden snail, 1.
Helix aspersa, French names for, 14.
Helix aspersa used in medicine, 3.
Helix hortensis, 16.
Helix ianthina, 129.
Helix ianthina found on the coast about Tyre and Beyroot, 129.
Helix lactea, 20.
Helix lactea cooked with rice, 20.
Helix lactea eaten in France and Spain, 20.
Helix lactea found in Corsica, 21.
Helix nemoralis, wood snail, 1.
Helix nemoralis found in Danish kjökkenmöddings, 3.
Helix nemoralis eaten at Toulouse, 16.
Helix Pisana, the banded snail, 2.
Helix Pisana, where found in Great Britain, 16.
Helix pomatia, apple or vine snail, 1.

Helix pomatia found in Kent, Surrey, Gloucestershire, etc., 2.
Helix pomatia, white variety and reversed specimens, 2.
Helix rhodostoma, 16.
Helix vermiculata, 17.
Hill of broken shells, 32.
Holland, the greatest supply of scallops is from, 100.
Holothuriadæ, 176.
Horse mussel (Mytilus modiolus), 52.
Horsewinkle, 135.
Hotel at Paris for pilgrims, 107.
Hydra of Lerna, a polypus or sepia, 170.

Ika-aurame, or squids, 172.
Illyrian snails mentioned by Pliny, said by Sir Gardner Wilkinson to be very numerous in Veglia or Veggia, 11.
Image of St. James, 110.
Incitatus, the favourite horse of the Emperor Caligula, 56.
Indian belief of the origin of pearls, 30.
Indians use Haliotidæ for ornaments, 114.
Ink of cuttlefish, 178.
Invention of oyster beds by Sergius Orata, 68.
Investigations of the Commissioners on the Irish fisheries, 75.
Irish names for cockle, 27.
Irish way of cooking cockles, 36.
Irish oysters, 74.
Irish pearls, 54.
Island of Ré and its oyster beds, 77.
Isle of Man scallop beds, 101.
Isocardia Cor, 42.
Isocardia Cor, account of, by the Rev. James Bulwer, 43.
Isocardia Cor, Mediterranean species, 44.
Italian names for the Pinna, 139.

Jacobitæ, or Jacobipetæ, 107.
Jacobsmuschel, 101.
Jambonneaux, 139.

Japanese pilgrims use the scallop-shell as a badge, 111.
Jardinière, 14.
Javanese belief that pearls breed, 30.
"Jemmy" the pearl-catcher, 54.
Jersey oysters, 73.
Jugurtha loses his treasures, 10.
Juice of the purple fish requires exposure to the sun to produce the colour, 128.

King John and the Milton fisheries, 73.
Kirkeens, or kirkeen thraws, 101.
Kjökkenmöddings, Danish, 33.
Kjökkenmöddings at Newhaven, Sussex, 33.
Kjökkenmöddings, Scotch, described by Rev. G. Gordon, 33.
Klaffmuschel, or Mya, 154.
Kraken, Norwegian, 169.
Kraken, altar erected on its surface, 170.
Kreaklingur, or mussel, 45.
Kunyu, or Mya truncata, 154.

Laborde, M., partakes of live snails, 7.
Lady's dress figured with dye of the purple fish, 129.
L'Aillado, 26.
L'Ayoli, or ail-y-oli, 26.
La Cacalaousada, 26.
Lana penna, 140.
Lana pesca, or fish wool, 140.
Land polypus mentioned by Pliny, 170.
Langskoel, 39.
Lapa burra, 113.
Lapa, or limpet, 121.
Large oysters mentioned by Pliny, 69.
Large oysters met with near Trincomalee by Sir James E. Tennent, 69.
Leaden coffins ornamented with scallop-shells, etc., 105.
Legend of St. James, 106.
Leigh oyster fisheries, near Southend, 71.

Leister, or trident, 166.
Leitrigens, to cook, 100.
Lepade, 121.
Lid scallop, 97.
Lid scallop used in shell-work, 97.
Lid scallop at Dawlish, 97.
Ligurian and snails, 10.
Limaia or limaio, 16.
Limaou and limat, 14.
Limassade, la, 26.
Lime made from calcined cockle-shells, 28.
Limpet, German names for, 121.
Limpet, habits of, 118.
Limpet juice and oatmeal, 119.
Limpet, or patella, 117.
Limpets, to dress, 122.
Limpets for bait, 118.
Limpets, large, on Devonshire coast, 118.
Limpets eaten at Eastbourne, 119.
Limpets, Eastbourne method of dressing, 122.
Limpets consumed at Larne, co. Antrim, 119.
Limpets eaten on the coast of Normandy, 119.
Limpet and oyster catcher, 121.
Limpets eaten at Plymouth, 119.
Limpets roasted, 122.
Limpet sauce, 122.
Limpet-shells found in cromlechs, 120.
Limpet-shells used for mortar, 121.
Limpet soup, 122.
Limpet soup at Naples, 119.
Limpets, large, in South America, 118.
Lincolnshire Fens supply Covent Garden with snails, 9.
Littorina littorea, 134.
Littorinidæ, 134.
Livrée, 16.
Loligo, or squid, 171.
Long oyster, or Pholas dactylus, 162.
Longherone, 45.
Lulu el Berberi, or Abyssinian oyster, 115.
Luma, and Gros Luma, name for Helix pomatia, 17.

o 2

Lustreless pearls, 61.
Lutraria elliptica, 153.
Lutraria maxima, or great clam, 101.
Lutraria oblonga, 153.

Mactra glauca, or helvacea, 152.
Mactra solida, 151.
Mactra stultorum, 151.
Mactra stultorum, roads made of the shells of, 152.
Mactra subtruncata, or lady cockle, 152.
Mactridæ, 151.
Mactridæ, to dress, 153.
Madre-perna, 139.
Madrid, price of oysters at, 81.
Manche de couteau, 41.
Marennes oysters, 79.
Meleagrina margaritifera, or white pearl-shell, 115.
Menestra de ostras y Almejas, 146.
Meninx, in Africa, famed for its purple, 131.
Messerschalenmuschel, 41.
Military order of Santiago de la Espada, 109.
Milk rendered luminous by a Pholas, 162.
Milton natives, 73.
Miranhá Indians, 117.
Mock asses'-milk, 6.
Mock pearls, 116.
Mogne, 16.
Mogul, anecdote of a, 59.
Moldavian snails, large, 11.
Molimorno, 16.
Moonbeams injurious to fish, 81.
Mossel, Dutch name for mussel, 45.
Mother-of-pearl made of Haliotidæ, 114.
Mother-of-pearl buttons, etc., 115.
Mother-of-pearl cups, 115.
Mother-of-pearl, crucifixes and beads made of, 115.
Mother-of-pearl, dishes and bowls of, 115.
Mother-of-pearl, fountayne and basen of, 116.
Mother-of-pearl, shippes made of, 116.

Mother-of-pearl, watch set in, 116.
Moucle de vigne, 17.
Moule, 45.
Mucianus and the oysters of Cyzicus, 69.
Muergo, Andalusian name for the razor-shell, 41.
Murex brandaris, 131.
Murex erinaceus destructive to oysters, 70.
Murex trunculus, 129.
Murex trunculus eaten in Spain, 132.
Muricidæ, 123.
Muschel, 45.
Muscl, muskel, muscule, Anglo-Saxon names for mussel, 45.
Muskels in brewet, 63.
Musselburgh, 48.
Mussel beds, or bouchots, 46.
Mussel, common, 44.
Mussels used for bait, 48.
Mussels and cockles, to cook, 67.
Mussels from Cornwall, etc., for Billingsgate, 47.
Mussels consumed at Edinburgh and Leith, amount of, 47.
Mussels and cockles in shell-mounds, 34.
Mussels dressed à la Provençale, 66.
Mussels, to dress, 65.
Mussels, large, from Exmouth, 51.
Mussels a valuable article of food, 51.
Mussels fit for food in the winter months, 52.
Mussels fed on spawn of starfish injurious to eat, 52.
Mussels, French trade in, 47.
Mussel fritters, 66.
Mussels injurious if gathered from ships' sides, etc., 51.
Mussel sauce, 65.
Mussels, scalloped, Francatelli's recipe, 66.
Mussels, seaweed, and shingle, render embankments firm, 49.
Mussels, little, called seeds, 46.
Mussel-shell for cutting the hair, 50.
Mussels sent to La Rochelle, 47.

INDEX. 197

Mussel soup, 65.
Mussel spawn, 45.
Mussels suspended from ropes, etc., attain a larger size than those which live on sand or mud, 47.
Mussels to be transplanted in July, 46.
Mussels, value of, in times of scarcity, 52.
Mutton-fish, or Haliotis Iris, 114.
Mya arenaria, 154.
Mya, natives of the Congo river collect a species of, 155.
Mya used for skimming milk, 29.
Mya, skin said to be poisonous of, 155.
Mya truncata, 153.
Myadæ, 153.
Myadæ, habits of, 154.
Myadæ, Hampshire method of cooking, 156.
Mye des Sables, or Mya arenaria, sold at Bordeaux, 155.
Mytilidæ, 44.
Mytilus edulis, 44.
Mytilus modiolus, 52.
Mytilus modiolus eaten in Ireland, 53.
Mytilus modiolus called the poisonous mussel at Tenby, 53.

Nacherone, 139.
Nahak, or rubbish collected by disease-makers in the island of Tanna, 125.
Napfmuschel, 121.
Napfschnecke, 121.
Napoleon I., the scabbard of his sword made of gold and mother-of-pearl, 116.
Nassa reticulata, 83.
Nassis, or osier kipe, 124.
Neapolitans eat mussels raw and fried, 66.
Necklaces of limpet and other shells found in British graves, 121.
Needle coated with copper, 79.
Nero's golden house, 115.
Neumann's description of the dog whelk, 127.

Newcastle glassmen, feast of the, 12.
Normandy oysters, 77.
Northumbrian oyster cultivation, 74.
"Nottle Tor," 34.
Nympsfield, 121.

Oatmeal and cockles, 36.
Octopi prized by the N. American Indians, 165.
Octopodia eaten by the modern Greeks, 165.
Octopods in market at Smyrna, 165.
Octopus vulgaris rare on British coast, 167.
Octopus vulgaris, specimens at Eastbourne and Babbicombe, 167.
Octopus vulgaris, French method of cooking, 175.
Odd method of cooking an oyster described by Evelyn, 84.
Œil de bouc, 121.
Oil of black snails, 7.
Old English rhyme on snails, 22.
Old pearls said to adhere to the shell, 55.
Olivette, or scallop, 99.
Ommastrephes, or flying squids, 172.
Onyches, 161.
Orders of knighthood which used the scallop-shell as an ornament, 109.
Orecchiale, 113.
Oriental pearls, 55.
Ormer, or ear-shell, 113.
Ormers fried or pickled in vinegar, 117.
Ormer-shells used to frighten birds from corn in Guernsey, 114.
Ormers, Jersey market supplied with, from the French coast, 113.
Ormer, to dress to perfection, 117.
Ormier, 113.
Ormond, 114.
Ostend oysters, 71.
Ostione, 80.

Ostras asadas, or fried oysters, 89.
Ostras en concha, scalloped oysters, 93.
Ostras en escabechados, pickled oysters, 95.
Ostras guisadas, ragoût of oysters, 90.
Ostras á la Pollada, 90.
Ostreadæ, 68.
Ostrea edulis, 68.
Ostrea Virginica at Cadiz, 80.
Ostrea Virginica on the French coast at Arcachon, 81.
Ostrea Virginica discovered by Sir G. Wilkinson at Tenby, 81.
Otaria, 114.
Otter-shell, Lutraria maxima, 101.
Oursin comestible, 178.
Oursin livide, 179.
Oursin melon, 179.
Ova, or Mytilus modiolus, 52.
Oxhorn cockle, 42.
Oxhorn cockles prized by the Brixham fishermen, 43.
Oyster, 68.
Oyster of Abydus, 70.
Oyster atlets, 87.
Oyster atlets curried, 87.
Oyster baskets in Paris, 80.
Oyster bed in Glenluce Bay, 76.
Oyster beds off Hayling and Portsmouth, 73.
Oyster bed in Lough Swilly, 75.
Oysters, charity, 61.
Oysters for consumptive people, 82.
Oysters from Cornwall, 74.
Oysters of Cyzicus, 69.
Oysters fattest at the full of the moon, 81.
Oyster forcemeat, 92.
Oysters, to fry, 89.
Oysters, fried, another way, 89.
Oyster fritters, 93.
Oysters au Gratin, 96.
Oysters, grilled, 90.
Oysters will not grow in the Baltic, 83.
Oyster heaps at Creggauns, in Tyrone, 83.

Oyster ketchup, 96.
Oyster loaves, 93.
Oysters, minced, 91.
Oysters, mussels, and periwinkles at Leigh, 71.
Oyster pie, 94.
Oyster powder, 95.
Oyster sauce, 86.
Oyster, brown sauce, 86.
Oyster sausages, to make, 91.
Oyster sausages, 91.
Oysters, scalloped in the old way, 92.
Oysters scalloped, 92.
Oysters always in season at New York, 81.
Oysters from Colchester sent to Leicester and Walsingham in Queen Elizabeth's reign, 71.
Oyster-shells for holy-water, 82.
Oyster-shell island on the east coast of Corsica, 82.
Oyster-shells as manure, 82.
Oyster-shells used by the Romans as toothpowder, 82.
Oyster soup, 84.
Oyster soup with fish stock, 85.
Oyster soup another way, 85.
Oyster trade in the vicinity of New York, 81.

Paignton method of cooking Cardium rusticum, 38.
Palostrega, 139.
Palourde, or Tapes, 144.
Palourde, or scallop, 101.
Pandore oysters, 76.
Parisians eat snails for breakfast, 19.
Patella athletica, 121.
Patella vulgata, 117.
Patelle, or limpet, 121.
Patellidæ, 117.
Patellidæ eaten by the ancients, 120.
Pearls called bones or stones by Greek authors, 55.
Pearls said to be congealed dewdrops, 30.
Pearl fishery on the coast of Columbia, 60.

INDEX.

Pearl fisheries of Condatchy, Aripo, and Manaar, 60.
Pearl fishery at Bahrein, 60.
Pearl fishery at Omagh, 54.
Pearls found in the Aplysia, or seahare, 53.
Pearls found in the oyster, scallop, cockle, periwinkle, etc., 53.
Pearls like black muscades, 58.
Pearl called the Sleeping Lion, 59.
Pearl, largest known, 60.
Pearl-lime, 61.
Pearl mussels in Lochs Earn, Tay, etc., 54.
Pearls in common mussel (Mytilus edulis), 53.
Pearl necklaces and chains for the hands and feet worn by the Persians and Medes, 55.
Pearls in Unio margaritiferus, 53.
Pearls preferred to other ornaments until the death of Maria Theresa, 57.
Pearl oyster, Meleagrina margaritifera, 55.
Pearl called la Peregrina, 59.
Pearl-shell snail, Turbo cornutus, 115.
Peasants near La Rochelle gather snails to send to America, 15.
Pecten Jacobæus, 103.
Pecten maximus, 100.
Pecten opercularis, 97.
Pecten varius, eaten in France, 99.
Peignes, 101.
Pelagia, the shellfish, 130.
Pelagium, the juice, or colour, 130.
Pellerinella, 99.
Periwinkle, 134.
Periwinkles mentioned by Athenæus, 135.
Periwinkles, to boil, 137.
Periwinkles in Brittany called Vrélin or Brélin, 135
Periwinkles, large consumption of, in London, 136.
Periwinkles of various colours, 135.
Periwinkles in kjökkenmöddings, 137.
Periwinkles abundant in Scotch kjökkenmöddings, 33.

Periwinkle, limpet, etc., found in the Irish oyster heaps, 83.
Periwinkles sent to London from Belfast, 136.
Periwinkles in the Orkneys, 136.
Periwinkles sent from Southampton to Jersey, 136.
Periwinkle soup, 137.
Periwinkles, to stew, 137.
Periwinkles called whelks at Belfast, 135.
Periwinkle, variety of form, 135.
Perles barroques, 57.
Petite palourde, or Pecten varius, 99.
Phasianella, or Venetian shells, 117.
Philoxenus the Solenist, 40.
Puoladidæ, 156.
Pholas used as bait, 161.
Pholas collected at Dieppe for food and bait, 163.
Pholas costata, a West Indian species, 163.
Pholas crispata, 163.
Pholas dactylus, 156.
Pholas, dried, recovers its luminosity when rubbed or moistened, 162.
Pholas sold in Jersey market ready boiled for eating, 163.
Pholas, Normandy method of cooking, 163.
Pholas, its perforating powers, a subject of discussion, 157.
Pholas, its phosphorescence, 162.
Pholas eaten raw in Spain, 163.
Pickled oysters, 94.
Pickled oysters for the London markets, Soyer's recipe, 95.
Pickling oysters in the Glamorganshire way, 94.
Piddock, or clam, 156.
Pilgrim offerings, 108.
Pilgrims-muschel, 101.
Pilgrim scallop, Pecten Jacobæus, 103.
Pincushions made of shells, 49.
Pinna, Aufrère describes the collecting of the, 140.
Pinna, British localities for the, 138.

Pinna forms a dish at an Attic banquet, 141.
Pinna at Dawlish, 138.
Pinna, or nacre, described by Pliny, 139.
Pinna a recherché dish at Naples, 141.
Pinna pectinata, 138.
Pinna soup, 142.
Pinna wool, 140.
Pinnæ fried like cutlets, 142.
Pinnophylax, 139.
Pinnoteres, 139.
Pinnotheres pisum, 51.
Pinnotheres veterum, 51.
Piscinola, the famous diver, 168.
Planorbis corneus yields a dye, 132.
Plato recommends the polypus to be boiled or roasted, 165.
Pliny and the luminosity of the Pholas after death, 161.
Pliny mentions several kinds of snails, 10.
Pliny recommends snails for a cough, 8.
Pliny's observations on the scallop, 97.
Pliny's supper, 11.
Poisoning by green oysters at Rochefort, 79.
Poli's method of cooking Cardium rusticum, 39.
Polpo, Italian name for the common poülp, 165.
Polynesian method of catching cuttlefish, 167.
Polypus said by Pliny to arrest hæmorrhage if bruised and applied, 170.
Polypus, its dangerous powers, 168.
Polypus, symbol of Messina, is figured on a medal of that city, 174.
Pontoppidan's description of the kraken, 169.
Pope, the, uses a purple robe to celebrate Mass in Lent and Advent, etc., 132.
Porphyra shellfish, discovery of the, 129.
Potage à la Poissonière, 85.

Poulp, habits of, 166.
Poulp in the Mediterranean, 165.
Poulp, large, at Nice, 167.
Poulps live in holes amongst rocks, 166.
Poultry fed on Patella vulgata, 121.
Poultry fed with lustreless pearls and grain, to restore brilliancy to the pearls, 61.
Powder-horns, 138.
Price of Haliotidæ in Channel Isles, 115.
Price of Helix aspersa, 17.
Price of Helix nemoralis, 16.
Price of Helix pisana at Marseilles, 16.
Price of Helix pomatia, 17.
Price of Helix vermiculata, 17.
Price of mussels taken at Lympstone, 47.
Price of oysters at Billingsgate in 1864, 80.
Price of scallops, 98.
Prices of Scotch pearls, 53.
Principal oyster beds, 70.
Professional pilgrim at Santiago de Compostella, 111.
Proper seasons for visiting Spain for scientific purposes, 19.
Proportions for mixing the juice of the buccinum and pelagium for dyeing wool, 130.
Protection to English pilgrims, 107.
Provençaux names for Helix aperta, 15.
Psammobia vespertina, or "the setting sun," 149.
Psammobia vespertina eaten at Kenmare, 149.
Psammobia vespertina, localities for, 150.
Pudworm, 161.
Puerto, Santa Maria, supplies Madrid with oysters, 80.
Pullers, sugar-loons, or colliers, 155.
Pullet, or cullyock, 142.
Pullet, or cullyock, used for bait, 143.

Purchase snails and eat them, 18.
Purple-dye used by the Nicaraguan Indians prepared from shellfish, 129.
Purple-dye produced from two kinds of fish, the Buccinum and the Purpura, purple or pelagia, 129.
Purple fish, 127.
Purple fish, various kinds mentioned by Athenæus, 130.
Purple imported from the Peloponnesus in the days of Ezekiel, 131.
Purpura Anglicana, 128.
Purpura lapillus, the dog-whelk, 127.
Purpura lapillus used for dyeing linen in Ireland in 1684, 127.
Purpura lapillus eaten in France, 132.
Purr, or butterfish, 143.
Pyrenean name, caracolo, for snails, 16.

Quadrans, a small copper coin, 11.
Quadrantes, 80, contained in a snail shell, 11.
Queen Elizabeth purchases Mary Queen of Scots' pearls, 57.
Queen Mary's parure of pearls, 58.
Queeus, or scallops, 99.

Radiata, or Echinodermata, 176.
Ragoût of snails, 12.
Ragoût of snails, Spanish recipe, 24.
Ragoût of oysters, 90.
Raw oysters beneficial to persons who suffer from weak digestions, 82.
Razor-fish on the Scotch coast, 42.
Razor-fish, to cook, 42.
Razor-fish soup, 41.
Razor-shell, or Solen, 39.
Razor-shells in the Bay of Concepcion, 41.
Razor-shells, collecting, 40.
Red Bank oyster-bed, 74.
Red whelk, almond whelk, Fusus antiquus, 132.

Red whelk used for bait at Dublin, 133.
Red whelk sold at Liverpool, taken on the Cheshire coast, 132.
Refuse heaps on the shores of the Moray Firth, 34.
Renouvelains, 46.
Rivers Irt and End, pearls found in, 30.
Roasted oysters, 90.
Rocher de Cancale oysters, 77.
Romans partial to snails, 10.
Roman ladies wore pearls at night, 56.
Romulus employed the purple dye for the trabea, 131.
Rossmässler and the empty snail-shells, 20.
Rufina, 101.
Ruocane, 27.
Rush baskets containing snails, 14.

Sabot, or periwinkle, 135.
Sacred geese in the temple of Juno, 10.
St. Clement's Church, Sandwich, 133.
St. James of Compostella performed many miracles, 108.
St. James, patron of Spain, 110.
Saintonge and Aunis, snails exported from, 15.
Salsa de Almejas, 146.
Sand clam, or Solen, 101.
Sauces for snails, 26.
Scalaria communis yields a purple liquor, 132.
Scallop great, Pecten maximus, 100.
Scallop called the butterfly of the ocean, 97.
Scallops, to cook, 113.
Scallop, its movement described by Mr. Gosse, 98.
Scallops, to dress, 112.
Scallops at Clavijo dropped there by St. James, 109.
Scallops, to fry, 99.
Scallops with matelote sauce, 99.
Scallop-shell in heraldry, 104.
Scallop-shell the badge of the pilgrim, 103.

Scallop-shell figured on coins, 111.
Scallop-shells used as lamps, 103.
Scallop-shells on monumental slabs, 105.
Scallop-shells belong legitimately to Compostella pilgrims, 107.
Scallops sent to the London market principally from Holland, 101.
Scallop soup, 99.
Scallops, to stew, 112.
Scallops at Vigo the constant food of all classes from Christmas to Easter, 101.
Scallops at Weymouth, 98.
Scarcity of Oxhorn cockle, 42.
Scheidenmuschel, 41.
Scotch kjökkenmöddings, 33.
Scotch pearls in demand abroad in twelfth century, 55.
Scotch pearl fishery revived, 53.
Scotch rivers contain pearl mussels, 54.
Scrobicularia piperata, or Mudhen, 143
Sea-birds feed on Patellidæ, 121.
Sea-egg, common, or sea-urchin, 176.
Sea-eggs sold in the market at Maracillea, 179.
Sea-eggs eaten raw in Chili, 178.
Sea-egg fishing in the Bay of Naples, 178.
Sea-snægl, or sea-snail, 135.
Season for oysters, 80.
Sea-urchin, anecdote of Lacedæmonian and the, 179.
Sea-urchin described by Pliny, 176.
Sea-urchin recommended to be eaten raw, 179.
Sea-wing, 138.
Sèche, Seiche, or Casseron, 164.
Seed pearls, 61.
Seeohr, 113.
Sepiadæ, 163.
Sepia, method of making, 173.
Sepia officinalis, 163.
Sepia used in painting, 173.
Sepiata, or supieta, 173.
Sepiola Rondeletti, 173.
Seppietta, or Calamaretto, 173.
Serranos, 19.
Serranos, stewed, 19.
Shannon oyster-beds, 74.
Shark-charmer, 61.
Shellfish good for those who take too much wine, 145.
Shellimidy, or snail, recommended for many diseases in Ireland, 5.
Shellimidy forragy, or periwinkle, 135.
Shell-lime, 49.
Shell-mounds of cockle-shells, 32.
Shell-mounds of St. Michel-en-L'Herm, 82.
Shells of Anodontæ used for skimming milk, 62.
Shells of Galicia, 107.
Shell-snails, pounded, for a swelling on the joints, 6.
Shells found in stone coffins, 108.
Shells used as trumpets in Muscovy and Lithuania by herdsmen, 125.
Shell trumpets used by sorcerers in the Island of Tanna, New Hebrides, 125.
Shell trumpets in Tahiti, 125.
Shelley, arms of Sir John, 133.
Shelly-meddings, 34.
Slip and Escallop-shell, order of the, 109.
Silesian way of feeding snails, 13.
Silkworm of the sea, 140.
Silver spoon boiled with mussels to prove if they are wholesome, 52.
Singular custom near Bordeaux, 13.
Sir J. E. Tennent mentions large oysters at Kottiar, near Trincomalee, 69.
Sir J. Gardner Wilkinson and the Illyrian snails mentioned by Pliny, 11.
Size of shell-mounds at St. Michel-en-l'Herm, 82.
Sliga-crechin, or the drinking shell, 29.
Sligane-mury, 101.
Small crabs in mussels said to make them unwholesome, 51.
Smirslingur, 154.

Smooth-shelled pearl mussels supposed not to contain pearls, 54.
Smurslin, 154.
Snail borne as arms in heraldry, 23.
Snail, crest of the Carpenters of Somersetshire, and of the Galay family, 23.
Snailery at Dijon, 17.
Snail feast at Newcastle no longer exists, 12.
Snail garden at Friburg, 12.
Snail garden in Lorraine surrounded with trellis-work, 12.
Snail hunters, 19.
Snail races, 21.
Snail, or shellimidy, 5.
Snail called Tardigrada domiporta, or the slow-going house-bearer, 22.
Snail, water, pectoral, 4.
Snails at Algiers sold in the market by the bushel, 21.
Snails cure ague, 3.
Snails a cure for asthma, 5.
Snails as bait for prawns, 21.
Snails at Cairo, 21.
Snails consumed in Burgundy, Champagne, etc., 17.
Snails for a consumption, 6.
Snails and earthworms for a consumption, 5.
Snails for a cough recommended by Pliny, 8.
Snails eaten in Corsica, 17.
Snails, small white, as a cosmetic, 9.
Snails used in the manufacture of cream, 9.
Snails exported from Crete, 17.
Snails eaten all the year round at Hyères, except at Easter, 15.
Snails for curing a web in the eye, 4.
Snails, to dress, 23.
Snails, to fatten, 13.
Snails fed on bran at Naples, 13.
Snails, fifteen species eaten on the Continent, 14.
Snails of woods and forests, 17.
Snails, grits of sand found in their

horns recommended for stopping toothache, 8.
Snails as food for birds, 9.
Snails cooked in the French way, 23.
Snails, on old French recipe for dressing, with a sauce, 23.
Snails give a flavour to wine, 19.
Snails at Hyères, 14.
Snails pounded for an impostume (whitlow), 8.
Snails for internal pains, 8.
Snails kept in jars, etc., 13.
Snails, large specimens from Moldavia, 11.
Snails as a medicine, 3.
Snails brought to Nantes on Sundays and fête days, 18.
Snails, Normandy way of cooking, another recipe, 24.
Snails sold in the Paris markets, 18.
Snails sent to Paris ready cooked, 18.
Snails formerly in Paris only found in the herbalists' shops, etc., 18.
Snails with parsley, Caracoles con Perejil, 23.
Snails, when poisonous, 13.
Snails as a plaster, 3.
Snails at the restaurants in Paris, 18.
Snails swallowed raw, a remedy for a weak chest, 8.
Snails, when considered in season in Paris, 19.
Snails and sheep's trotters for a consumption, 6.
Snails for sheep, said to flavour the mutton, 9.
Snails smoked and dried, 26.
Snails from Soletum, 11.
Snails and black soap, a cure for corns, 3.
Snails eaten in Spain not only by the poorer classes, 19.
Snails, Spanish method of eating, 19.
Snails, stuffed, considered very good, 26.
Snails eaten in Syria, 21.

Snails, method of transporting live, 15.
Snails, or escargots, kept in winter by the vine growers of Dijon in trenches dug in the vine slopes, 13.
Snails at Vienna, 13.
Snail-shells, ashes of, good for the gums, 8.
Snail-shells found at Auch, Agen, etc., 3.
Snail-shells found in kjökkenmöddings, 3.
Snail-shells found at Lymne, in Kent, 3.
Snail-shells found on the sites of Roman stations, 2.
Snails-shells holding 40 sixpences, 11.
Solen, or razor-shell, 39.
Solen ensis, 39.
Solen ensis eaten in the Feroe Isles, 39.
Solenidæ, 39.
Solenist, Philoxenus called the, 40.
Solenistæ, people so called who collected solens, 40.
Solen marginatus, or vagina, 39.
Solen marginatus prized as an article of food by the Neapolitans, 39.
Solen siliqua the largest British species, 39.
Solens an expensive dish at Naples, 41.
Solens prized in Japan, 41.
Solens mentioned by Ulloa, 41.
Solens, another way to cook, 42.
Sopa de Almejas, or Tapes soup, 145.
Soyer's recipe for cooking mussels, 64.
Soyer's recipe for pickling oysters for the London markets, 95.
Soyer's method of cooking scallops, 99.
Soyer's porridge of cockles, 35.
Spaniards hand white wine round with shellfish, 145.
Spanish cure for consumption, oil of black snails, 7.

Spanish cure for the headache, 7.
Spanish way of making fish sauce, 150.
Spanish method of cooking all kinds of shellfish, 151.
Spanish recipes for cooking snails with rice, butter, etc., 20.
Spout-fishes, 39.
Springing Loligo mentioned by Pliny, 172.
Squid highly esteemed by the ancients, 171.
Squid or squill used for bait, 171.
Squid-fishing in Japan, 172.
Squid, or calmar, eaten on the French coast, 171.
Squinns, 99.
Starfish feeds on oysters, 70.
Steam fishing vessel built at Cockenzie, 76.
Steckmuschel, 139.
Stumpfmuschel, 150.
Sugar-loons, 155.
Sun, the setting, or Psammobia vespertina, 149.
Superiority of British oysters, 68.
Superstitions of the Ceylonese divers, 61.
Superstitions of the Scotch fishermen, 76.
Superstitious dread of freshwater mussels, 63.
Syrup of snails, 7.

Tapa, tapada, or tapet, names for Helix aperta, 15.
Tapes, or Almejas, 145.
Tapes aurea eaten in Ireland, 144.
Tapes aurea found in the Scilly Isles, 145.
Tapes cooked another way, Almejas cocidas, 146.
Tapes cooked Hampshire method, 147.
Tapes decussata, Almejas blancas, 145.
Tapes decussata eaten in Devonshire, Hampshire, and Sussex, 143.
Tapes decussata more local than Tapes pullastra, 143.

Tapes decussata common near Exmouth, 143.
Tapes decussata, called purr, and butterfish, 143.
Tapes decussata, how to find, 144.
Tapes decussata called Clouvisso or Clovisse on the French coast, 144.
Tapes decussata highly prized by the Spaniards, 144.
Tapes au naturel, Almejas al natural, 146.
Tapea, potage of oysters and, Menestra de Ostras y Almejas, 146.
Tapes pullastra, pullet or cullyock, 142.
Tapes pullastra, a common species, 143.
Tapes pullastra used for bait in the Northern Isles, 143.
Tapes ragoût, Almejas guisadas, 146.
Tapes sauce, Salsa de Almejas, 146.
Tapes soup, Sopa de Almejas, 145.
Tapes Virginea varies much in colour, 144.
Tapes Virginea at Dawlish and Tenby, 144.
Taprobane, island of, most productive of pearls, 55.
Tarentine, red, 131.
Tarentum, ancient dyeing-houses at, 131.
Tavernier's pearls, 60.
Tellinidæ, 149.
Tellinidæ rarely used for food in Great Britain, 149.
Tellinidæ mentioned by Athenæus, 150.
Tellinidæ, sauces made of. 150.
Teredo, account of the, 157.
Teredo said to be good to eat, and excelling all shellfish, 160.
Teredo navalis and Teredo norvegica, 160.
Teuthidæ, 173.
Teuthis, Aristotle speaks of the, which has ink of a pale colour, 173.

Theognis, riddle of, 127.
Theophrastus on the habits of snails, 16.
Thrushes partial to snails, 10.
Tootoofe, 167.
Torbay-noses, or Oxhorn cockles, 43.
Torbay-noses, to dress, 44.
Trabea, Romulus uses the purple dye for the, which was purple and white, 132.
Trabea, Servius mentions two other kinds of, 132.
Trabea, the royal robe worn by the early kings, 132.
Trade, oyster, with Belgium, 71.
Trade, pickled oyster, between London and Glamorganshire, 75.
Trade in snails at Covent Garden, 9.
Tridacna gigas, shells of, used for holy-water, 82.
Trigonia pectinata, an Australian bivalve, 117.
Trigonia pectinata, bracelets, brooches, etc., made of the shells, 117.
Trochidæ sold occasionally as winkles at Jersey, 136.
Trochus found in the Creggaun heap, with the shells of the oyster, mussel, etc., 83.
Trogmuscheln, 152.
Trompetenschnecke, or Kinkhorn, 127.
Trough-shell, or Mactra, 151.
Troyes supplies Paris with the apple or vine snails ready boiled in their shells, 18.
Tumps, 121.
Turbinellidæ, 126.
Turbinella rapa, or chank shell, 126.
Turbinella rapa as a wind instrument, 126.
Turbinella rapa sawn into rings for anklets and bracelets, 126.
Turbinella, reversed shells of, highly prized by the Chinese, 126.

Turbinella, consecrated oil kept in reversed shells of, by the Chinese, 126.
Turbinella rapa, or sacred shell of the Buddhists, 126.
Turbo cornutus, the snail pearl shell, 115.
Tympana, or hand-drums of the ancients, 56.
Tympania, or tambour-pearls, 55.
Tyre and Beyroot, Helix ianthina common on the coast about, 129.
Tyre, the purple of, the best in Asia, 130.
Tyre, said by Strabo to have had numerous dyeing works, rendering the city unpleasant as a place of residence, 130.
Tyre, holes observed there by a modern traveller, Mr. Wilde, cut in the solid sandstone rock in which shells seem to have been crushed, in ancient times, 131.
Tyrian medals, 130.
Tyrian purple hue given to wool by soaking in the juice of the Pelagia, and afterwards dipping it in the juice of the Buccinum, 130.
Tyrian purple said to have been produced from Murex brandaris, 131.

Ulm, celebrated for its escargotières, 13.
Unionidæ eaten in the south of Europe, 62.
Unionidæ, roasted in their shells, or drenched with oil, etc., 62.
Unionidæ and Anodontæ used for bait in the neighbourhood of Nantes, 62.
Unio margaritiferus, freshwater pearl-mussel, 53.
Unio Requienii and Unio litoralis found near Granada in the river Jenil and brought to the market, 62.
Unio tumidus and Unio pictorum produce small pearls, 62.

Value of Mary, Queen of Scots' pearls, 58.
Vanneau, or olivette, 99.
Various shells called clams, 101.
Veghia, or Veggia, the Cyractica of Strabo, 11.
Veneridæ, 142.
Venus Chione, or Cytherea Chione, 148.
Venus Chione recommended by Poli as most excellent food, 148.
Venus Chione found at Plymouth, etc., 148.
Venus mercenaria, the clam, 102.
Venus verrucosa, or the Warty Venus, 147.
Venus verrucosa sold in the market at Algiers, 147.
Venus verrucosa found in the English Channel, Channel Isles, etc., 147.
Venus verrucosa, to cook, 149.
Venus verrucosa collected at Herm, near Guernsey, for food, 147.
Venus verrucosa eaten at Kenmare, Ireland, and also in county Clare, 147.
Venus verrucosa cultivated on the coast of Provence, 148.
Veasels called "cogs," short and of great breadth, like a cockleshell, whence the derivation of the name, 34.
Vignot, vignette, French names for periwinkle, 135.
Village of Charron, a large mussel trade at the, 45.
Vinaigrette, la, a sauce for snails, 26.
Vrélin, or brélin, the periwinkle in Brittany called, 135.

Walton, the Irishman, first established mussel beds on the French coast, 45.
Wampum, or Indian money, 102.
Wampum made of the clam Venus mercenaria, 102.
Wampum, the, token of peace amongst the American Indians, 102.

INDEX.

Water rats, and Dreissena polymorpha, 62.
Wedge-shell, or Donax, 150.
Weight of mussels sent at a time to Billingsgate market, 47.
Welsh rivers contain pearl-mussels, 54.
Weolc, whelk in Anglo-Saxon is, 127.
Weolc-basn-hewen, meaning of, 127.
Weolc-read, or scarlet dye, 127.
Wexford oyster-beds, 75.
Wexford oysters taken to the French coast for laying down, 75.
Whelk, buckie, or conch, 124.
Whelk, an enemy to other mollusks, 123.
Whelk, or purpura, symbol of the city of Tyre, appears on the Tyrian medals, 130.
Whelk sculptured on font in St. Clement's Church, Sandwich, 133.
Whelk soup, 134.
Whelk soup, another way of making, 134.
Whelk-tingle, or sting-winkle, 124.
Whelk, a species of, used as trumpets in North Wales by the farmers for calling their labourers, 125.
Whelk, white varieties of the red or almond, 133.
Whelks for bait, 124.
Whelks supplied to Billingsgate chiefly from Harwich and Hull, 124.

Whelks taken in wicker baskets, 124.
Whelks, to dress, 134.
Whelks, Dublin method of cooking, 134.
Whelks borne in heraldry, 133.
Whelks feed on oysters, 70.
Whelks, season for, 124.
Whelks troublesome to lobster-fishers, 124.
White bones or pearls, 55.
White oysters from Spain, Bretagne, etc., sent to Marennes, 79.
White oyster sauce, 86.
White snails from Rieti, 10.
Whitstable oyster beds, 72.
Whitstable a fishing town of note in the reign of Henry VIII., 71.
Whistles made of the shells of Helix pomatia, 21.
Wigwam Cove, piles of old shells at, 32.
Winter soup of snails, 24.
Witch goes to sea in a mussel shell, 49.
Women of the Shir tribe make girdles and necklaces of river mussel-shells, 117.
Wood snail, Helix nemoralis, 1.
Wordsworth's lines on the limpet, 118.

Yoags, 53.
Youghal way of cooking sugar-loons, 156.

Zostera marina, 179.
Zots-kappen, 43.

THE END.

PRINTED BY J. E. TAYLOR AND CO., LITTLE QUEEN STREET, HOLBORN.

M L del _ G B Sowerby, lith. Vincent Brooks, Imp

1. Cardium edule _ Common cockle
2. Cardium rusticum _ Red nose cockle.

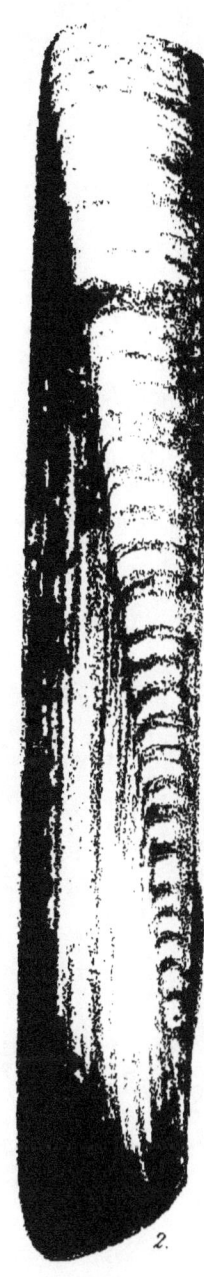

M. del. — G. B. Sowerby, lith.   Vincent Brooks, Imp

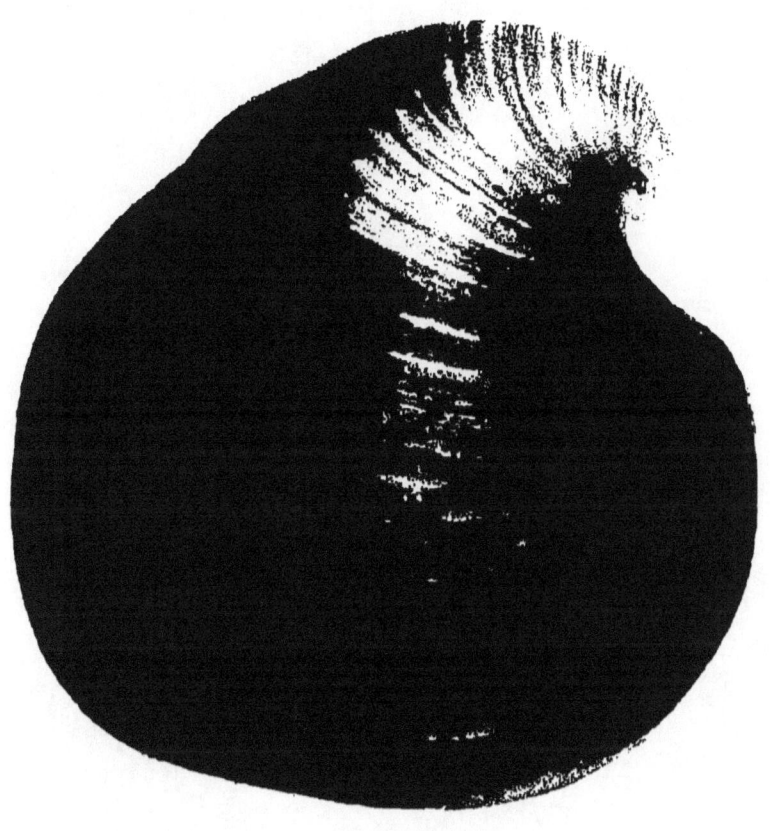

M del   G. B Sowerby, lith   Vincent Brooks, Imp

Isocardia Cor — Heart shell or Oxhorn Cockle

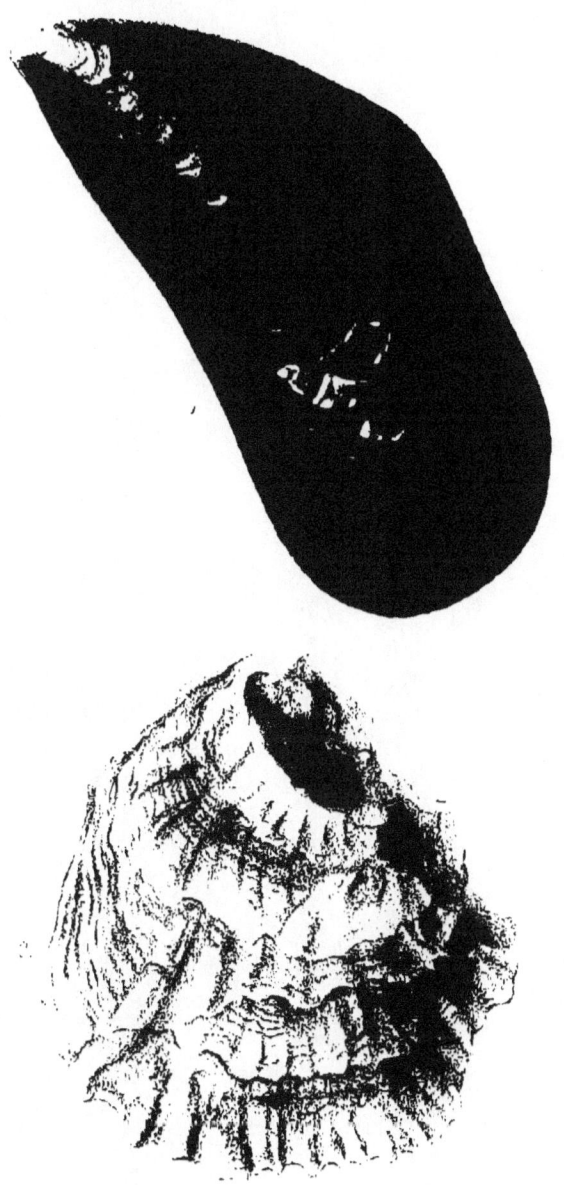

del. G B Sowerby, lith.   Vincent Brooks, Imp

1. Mytilus edulis. Common Mussel.
2. Ostrea edulis. Oyster.

6.

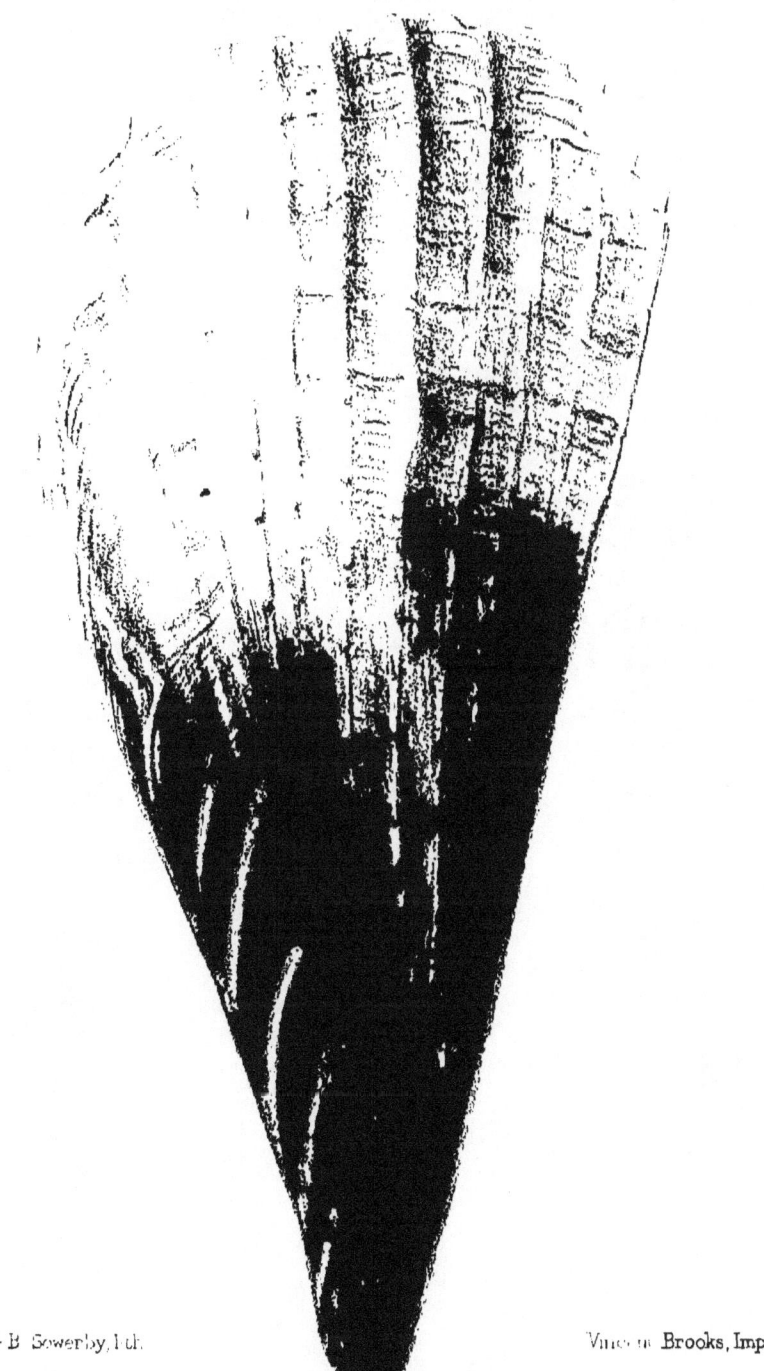

del — G B Sowerby, lith.   Vincent Brooks, Imp.

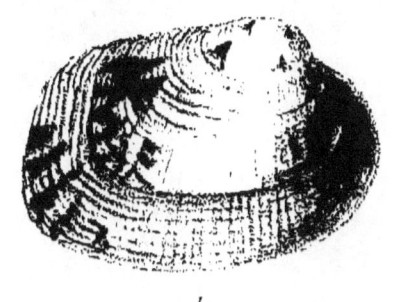

G. B Sowerby, lith.   Vincent Brooks, Imp

1 Tapes pullustra. Pullet.
2. Venus Verrucosa. Warty Venus.

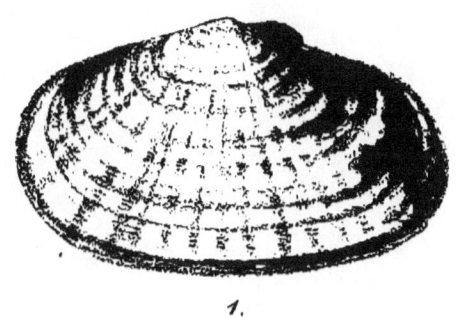

del _ G B Sowerby, lith   Vincent Brooks, Imp.

1 Psammobia Vespertina   The setting Sun
2 Mactra Solida, or Trough shell

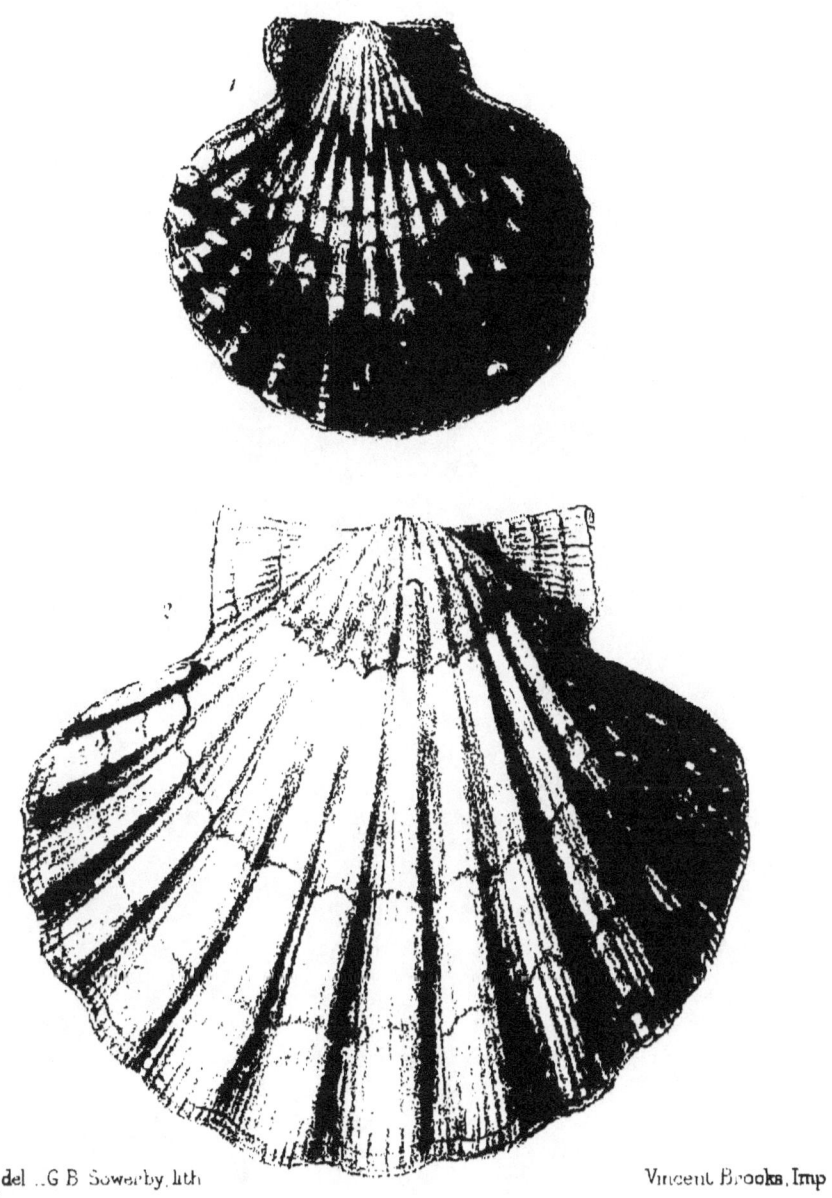

1 Pecten Opercularis or Painted scallop.

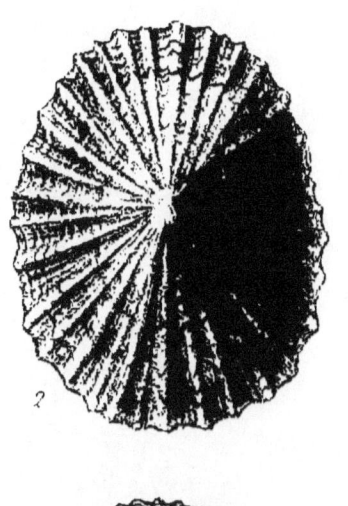

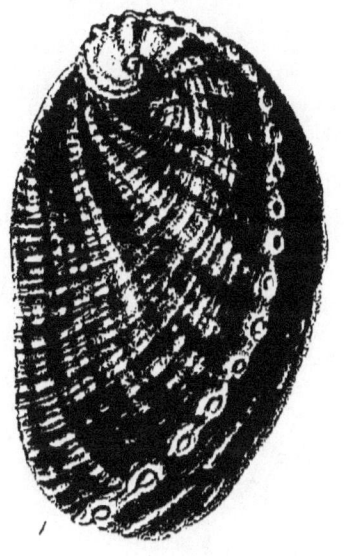

del_G. B. Sowerby   Vincent Brooks, Imp

1 Haliotis tuberculata, Ear-shell, or Sea-Ear.

1 Buccinum undatum Whelk

12.

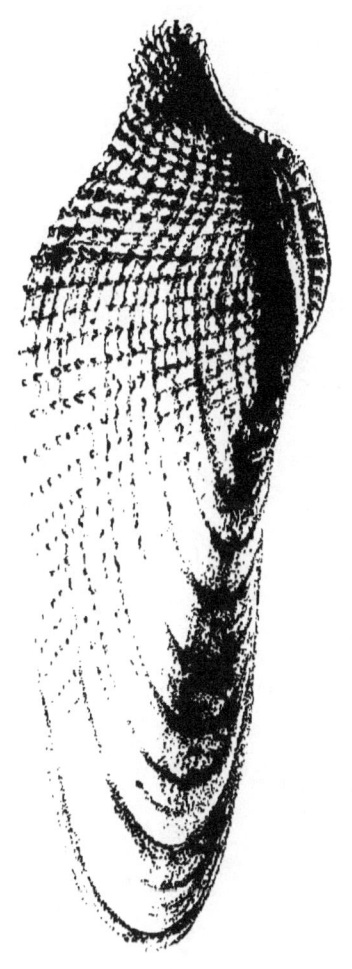

Pholas Dactylus. Piddock or Clam

www.ingramcontent.com/pod-product-compliance
Lightning Source LLC
Chambersburg PA
CBHW021353230426
43666CB00006B/512